T0198368

Marshmallow Math

Early Math For
Young Children

A fun and novel way to teach your child
the fundamentals of math.

Trevor Schindeler

Puts the fun back into fundamentals!

Order this book online at www.trafford.com
or email orders@trafford.com

Most Trafford titles are also available at major online book retailers.

Written and illustrated by Trevor Schindeler.
Layout by Rob Brownlee.

Note for Librarians: A cataloguing record for this book is available from Library
and Archives Canada at www.collectionscanada.ca/amicus/index-e.html

Printed in Victoria, BC, Canada.

ISBN: 978-1-5539-5395-1 (Soft)
ISBN: 978-1-4122-5129-7 (e-book)

*We at Trafford believe that it is the responsibility of us all, as both individuals and corporations,
to make choices that are environmentally and socially sound. You, in turn, are supporting this
responsible conduct each time you purchase a Trafford book, or make use of our publishing services.
To find out how you are helping, please visit www.trafford.com/responsiblepublishing.html*

*Our mission is to efficiently provide the world's finest, most comprehensive book publishing
service, enabling every author to experience success. To find out how to publish your book, your
way, and have it available worldwide, visit us online at www.trafford.com*

 www.trafford.com

North America & international
toll-free: 1 888 232 4444 (USA & Canada)
phone: 250 383 6864 ♦ fax: 812 355 4082 ♦ email: info@trafford.com

Dedicated to Ryan and Tamara whose
love of learning inspired me to write this book.

Table of Contents

CHAPTER ZERO

Introduction

Why Math is Important

It is more important than ever to be good at working with numbers. While a generation ago it may have been possible to be successful in life with minimal math skills, today most occupations require good math skills. It is sometimes said that the invention of calculators has made it unnecessary to be good at math. In fact, the opposite is closer to the truth. The proliferation of calculators reflects a need to understand math that is greater than ever. The reason why we have so many calculators around is because math has become such an integral part of our lives. However, calculators are of little assistance to someone who does not understand math. Knowledge of math is still necessary in order to identify problems, analyze data, and determine solutions.

Purpose

The purpose of this book is to help you teach your child the fundamentals of math. The approach set out in this book will help your child to develop a sound understanding of basic number concepts including counting, addition, subtraction, multiplication, and division. If your child truly understands the fundamentals and has mastered basic math skills, he or she will have both the capacity and the confidence to excel at learning more advanced mathematics.

Learning Math Is Fun

Children have an innate desire to learn and they love to learn math as much as any other subject. Children enjoy learning to count, add, and subtract as much as they enjoy learning their ABC's. Math can be fun and should be as much a part of every childhood as nursery rhymes and bedtime stories.

Overview

This book is divided into four parts. Part One focuses on counting and introduces the concepts of addition and subtraction. Part Two explores mathematical skills that do not involve arithmetic such as sorting and comparing, telling time, spatial awareness, pattern making, and elementary geometry. Part Two also suggests activities that will help your child learn to read and write numbers. Part Three continues to develop counting, addition, and subtraction skills. Part Four introduces several number skills, such as multiplication and division, which go beyond what is ordinarily expected of children at a grade one level.

Not Traditional
Exercises

Young children may not recognize the written symbols, called "numerals", that we use to represent numbers and can have great difficulty printing them. However, children can learn important number concepts long before they can read and write numbers. It is, therefore, a mistake to wait to introduce math until your child can read and write numbers. For this reason, most of the suggested activities in this book do not involve written work or traditional exercises. As well, since this book does not rely upon written work or traditional exercises, the activities can be done virtually anywhere and at anytime. You can do many of them while waiting for a bus, driving in the car, or even playing in the backyard. Furthermore, it is best to repeat most activities over and over and over. It is easier to do so if you are not relying on traditional exercises. Lastly, by emphasizing mental math over written work, your child's ability to manipulate numbers in his or her head will be strengthened. This will provide a strong foundation for future math skills.

Rote
Memorization

The concept of "rote memorization" has often been criticized. In the absence of understanding, rote memory is of little value. However, a good memory for number facts together with a firm understanding of the underlying concepts results in exceptional math skills. We tend to overlook that our "knowledge" of certain number facts, such as that "five plus five equals ten", is based upon both our understanding of the concept of addition and our memory. We know what it means to add two numbers together and we have memorized certain common equations. Once understanding is solid, the more number facts your child memorizes, the more proficient he or she will be at math. Even in the age of calculators and computers, memorizing number facts will help your child to better comprehend problems, recognize patterns, and develop solutions.

Counting Objects

Many of the activities described in this book involve having your child count out small objects such as marshmallows. Counting objects are sometimes called "manipulatives", something that your child can pick up and handle. Marshmallows can be substituted with apples, oranges, wooden blocks, toys, pennies, jellybeans, dry macaroni noodles, and the like. Having physical objects such as fingers and toes to look at and count helps to make abstract concepts more concrete for children. Marshmallows are simply more numerous than fingers and toes and much easier to place into piles! If you have a very young child, do not give him or her small objects to count because of the danger of choking.

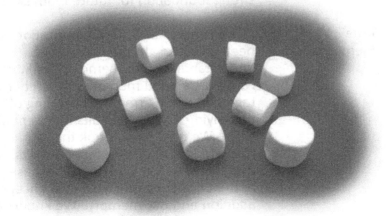

Play Time

Think of the activities included in this book as a form of play. Learning to count, add, and subtract should always be fun. If you take a positive approach, your child will feel eager and excited about doing the activities. It is advisable to spend only short periods of time doing math with your child. Five to ten minutes with a young child, and perhaps ten to fifteen minutes with an older child, is sufficient. Choose a time of day when your child is usually bright and active. Once your child appears to be tired or distracted it is time to stop.

One-on-one

If at all possible, do the activities with your child while working together one-on-one. Generally speaking, your child will not be able to do the activities on his or her own. Likewise, they are not activities that can be done with a large group of children such as in a classroom setting. Working one-on-one with your child will allow you and your child to achieve great success even though you only do math for very short periods at a time.

Focused Attention

Short periods of focused attention from both yourself and your child can result in remarkable achievement. Your child will thrive on the focused attention that you are giving and is likely to respond in kind. While your child is focusing hard on learning, you will have to focus on providing him or her with suitable exercises and questions. You will also have to focus on your child's answers and provide assistance whenever required. Each time that your child completes an activity or answers a question consider a number of things. Has your child mastered the concept? Was the question too easy or too difficult? Should the next question be more or less challenging? Should we start learning a new skill or reinforce a skill that was learned earlier? Your objective is, at one and the same time, to both challenge your child and to ensure his or her success.

Jump Around

Try to incorporate a little math into your daily routine. If counting and using numbers become a regular activity, you will notice remarkable progress. The activities are not meant to be done only once and then forgotten. Repeat activities until your child masters them and occasionally revisit skills and activities that were mastered earlier. In fact, the best approach may be to jump all around spending only a few minutes on any one skill or activity. This will keep both your interest level and your child's interest level high.

Learning Materials

It will be helpful to keep a small collection of learning materials handy. To start with, keep a plastic container of suitable counting objects. For toddlers, it may be marshmallows, wooden blocks, or other toys. For older children it could be dry macaroni noodles, pennies, or jellybeans. Over time you will add other learning materials to your collection including a small container of coins, several pairs of dice, ruler, tape measure, fraction circles, tangrams, geometrical shapes, pencil, paper, and a good eraser.

You're the expert

Many of us did not do well in math. To some extent our parents and the school system may have failed us. You may be reading this book because you did poorly in math and want your child to do better. You may even be apprehensive about your ability to help your child learn math. Please rest assured that if you can count, add, and subtract you will have no trouble helping your child to learn the fundamentals of math. Remember that, from your child's point of view, you will be an expert, a veritable walking, talking calculator. The fact that you failed algebra will not be an issue. Simply follow the advice of good trial lawyers; never ask a question that you do not know the answer to. Your efforts now, with the easy stuff, will help your child gain the ability and confidence to excel at doing the hard stuff in later years.

Intrinsic Motivation

Children are highly motivated to learn. The motivation is intrinsic. It is part of their essential nature to want to learn and develop skills and children receive enormous personal satisfaction from doing so. At the same time, children desire recognition and acknowledgement of their learning from their parents. It is important, therefore, to provide your child with the recognition and acknowledgement that he or she desires. Celebrate your child's learning with them. However, try to avoid the temptation of providing your child with other rewards for learning. Giving your child such things as stickers, stars, or treats places the emphasis on external motivators. In time, such external motivators may displace your child's intrinsic motivation. Your child may come to see learning merely as a way to receive a reward and lose the sense of deep personal satisfaction that learning can bring.

Thinking Mathematically

You will know that your efforts have been rewarded when numbers become a part of your child's life. That is, when number concepts help your child to better understand the world around them and he or she begins to use number concepts to communicate his or her own thoughts and ideas. That is, when your child begins thinking mathematically.

Off and Counting ONE

Part One focuses on counting and introduces the concepts of addition and subtraction. The chapters in Part One follow a natural and gradual progression of skill development, with each new skill building upon a foundation of earlier skills. However, it is recommended that the reader also explore Part Two at the same time. Part Two discusses important early mathematical concepts and skills, such as pattern making and spatial awareness, which do not fit neatly within an orderly progression of skills. The skills discussed in Part Two may be developed at the same time as the skills discussed in Part One.

CHAPTER ONE

Counting to Ten

1

Counting is fundamental to a child's understanding of numbers. It is probably never too early to start counting with your child. Count fingers, toes, toys, books, apples, family members, pets, etc. As ten is a significant number in our "base ten" number system, ten is an appropriate point at which to stop counting and start again.

Some children learn to recite the number names "one, two, three, four, five, six, seven, eight, nine, ten" like a poem, without understanding the concept of number or quantity. A young child who rhymes off "one, two, three, four" all the way up to ten like a pro may not have any idea what "four" means or what "seven" means. That is why it is a good idea to count out objects of some sort rather than to just count aloud.

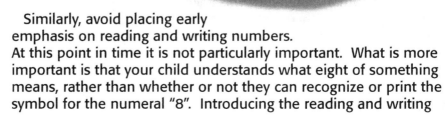

Similarly, avoid placing early emphasis on reading and writing numbers. At this point in time it is not particularly important. What is more important is that your child understands what eight of something means, rather than whether or not they can recognize or print the symbol for the numeral "8". Introducing the reading and writing of the symbols too early may confuse the child about the meaning of numbers. This is different from learning the alphabet where symbol recognition should be encouraged from the start. With the alphabet, unlike numbers, the primary concept is recognizing the letter symbols and knowing the sounds that they represent. See Chapter 28 for a discussion about when and how to introduce the reading and writing of numbers.

Start off by giving your child a small number of objects to count, perhaps only three or four. Please remember that noodles and other small objects may not be appropriate for very young children because of a choking hazard. Use a "choke tube" to ensure that any small counting objects are safe for your child. A choke tube may be purchased at any good children's store and is an excellent safety device to own if you have a toddler in the house. Try counting apples and oranges or teddy bears.

Once your child is able to count a small number of objects with confidence, slowly provide him or her with a larger number of objects to count. Always be prepared to assist your child in counting.

Even after your child can count all the way to ten, it is a good idea to occasionally count to a smaller number. Put out various numbers of objects to be counted and change the number randomly. This will help to ensure that your child really grasps the concept of quantity and is not just reciting the words.

Encourage your child to do different things with the objects that he or she is counting. A few ideas are to put them in a row, put them in a dish, make a new pile, or hide them. This will help to sustain your child's interest level.

Children love to count and will have fun doing so over and over. Remember that repetition is good and will reinforce your child's understanding of numbers and establish a foundation for future math skills. It is repetition that will stimulate your child's brain to grow and reinforce the neural pathways that process mathematical concepts.

Depending on the age of your child, it may be months or years before he or she moves beyond this stage in their competency with numbers. Learning to count, like learning to talk, is an enormously difficult thing to do. It is far more difficult than most of the things that adults are called upon to learn, so don't be in too big of a hurry to move on.

CHAPTER TWO

One-to-One Correspondence

2

One-to-one correspondence is a mathematical way of saying that if there are six people and you want to give each person a cookie, you will need six cookies. This is a simple, but fundamental, concept that children must learn. It is a small, but important step, from simple counting. The above illustration using cookies is a good example of how to introduce the concept to your child. If you have a large family, then every meal presents a good opportunity to reinforce the concept. How many plates will we need? How many glasses will we need? How many knives and how many forks? At snack time, how many apples or oranges will we need? If you have a small family and especially if there are only the two of you, you will want to improvise. Throw a tea party for teddy bears or take toy animals on a picnic in the living room. If you have friends or family come to visit, take advantage of the opportunity to reinforce the concept with a different number of people.

With older children, you may also explore two-to-one and three-to-one correspondence. This will provide a foundation for understanding multiplication later on.

CHAPTER THREE

I'm Thinking of a Number

3

This little game helps young children to learn the order of numbers without always starting at one and reinforces the concept that numbers change in size or magnitude. It also encourages logical thinking.

Simply choose a secret number between one and ten and then ask your child to guess what number you have chosen. If your child guesses the wrong number, let him or her know whether the number you are thinking about is smaller or bigger than the number they suggested. Repeat the process until they guess correctly. Try changing roles and attempt to guess the secret number chosen by your child.

Once your child knows how to read numbers, you might want to use a number line to play this game. Print out the numbers from one to ten on a piece of paper. Have your child circle numbers that are too small and cross out numbers that are too big. This will help your child to pin down the secret number.

MARSHMALLOW MATH

1 2 3 4 5 6 7 8 9 10

    Remember to also play this game when your child is learning larger numbers, especially numbers ten to twenty. The odd names assigned to such numbers make them particularly difficult to recall.

Adding to Ten by One's

4

Once your child has grown skilful counting to ten, it is a good time to introduce the concept of addition. Counting is really just a form of addition, adding by one each time. The next step in developing your child's understanding of numbers is to make this more explicit.

Very simply, when counting to ten, add words such as "plus one equals" or "plus one more makes" and then allow your child to supply the answer. For example; "one plus one equals...?", "two plus one more makes...?", "three plus another one equals...?", etc.

You may be surprised to find that your child has to stop and think about the answer. While it may seem to be a minor change in approach, the concept of addition introduces a new element to the activity. It is moving further away from reciting the numbers from one to ten like a poem. Adding the words "plus one equals" and waiting for a response breaks up the rhythm of simple counting and makes your child have to think about the answer. Be prepared to help with the answer if required.

MARSHMALLOW MATH

It is, of course, important to continue using objects, such as marshmallows, pennies, or jellybeans to illustrate the point. Encourage your child to count out all of the objects before giving an answer. This will interrupt simple counting and make your child think more about the process. The growing number of objects will demonstrate the concept of increasing quantity.

Avoid always starting this activity at number one. Set out different numbers of objects to which your child can add one more.

CHAPTER FIVE

Counting Backwards from Ten

5

As the saying goes, a real expert knows the subject "forwards and backwards". Counting backwards isn't just showing off or something you do when launching rockets! It is the basis for understanding subtraction. It also demonstrates to your child that numbers work in both directions and that there is order and consistency.

Counting backwards will probably seem difficult for your child at first compared to counting forwards, which will seem so easy now. It is easy to forget the hundreds of times that you counted with your child before he or she could count to ten on their own. Chances are that you will not count backwards with your child nearly as many times as you counted forwards before he or she masters it. Counting backwards exercises your child's memory big time. Think about it. How quickly could you, for example, count backwards from fifty by sevens?

In ordinary life there are fewer reasons, considered here to be opportunities, to count backwards than there are to count forwards. Unless you happen to work for NASA, it isn't often that we are called upon to count backwards. It is possible, however,

MARSHMALLOW MATH

to create opportunities. Instead of just counting out how many knives you are placing on the table, decide on how many you need and then count backwards as they are set out on the table. At the store, decide on how many apples or oranges you need and then count backwards to zero as you place them in a bag.

CHAPTER SIX

Counting to Twenty

Chances are good that your child has already started counting past ten. Children have an innate desire to learn, to develop their skills and abilities, and to push themselves further and further. Take advantage of your child's enthusiasm to advance his or her knowledge of numbers.

Counting to ten makes a good first milestone in the long journey of learning to count well. Twenty makes a good second milestone to aim for. This is partly because, being a multiple of ten, it is a nice round number. The real reason it makes such a good milestone, however, is because the numbers between ten and twenty are so confusing. Eleven, twelve, thirteen? Who made up such strange numeral names that do not even repeat for another hundred numbers? Children can find remembering these numbers names difficult. Fortunately, you may find that once your child can count to twenty with confidence, learning to count beyond twenty will be easier.

Helping your child learn to count to twenty will be similar to helping your child learn to count to ten. Continue counting out objects to reinforce the sense of quantity. At this point you will probably only count smaller objects with your child such as macaroni noodles, pennies, or jellybeans as opposed to apples or family members unless, of course, you have a really big family. However, it is fun to count larger things when the opportunity presents itself such as houses on a street, cars passing on the road, or cows standing in a field. You will find that children are happy to count whenever prompted.

MARSHMALLOW MATH

Unless your child is already learning to print and is asking about how to read and write numbers, it is probably still too early to introduce these elements into the activity. Developing a sense of numbers and basic math skills has very little to do with recognizing numerals. As discussed previously, what young minds have to concentrate on are quantity, order, and the relationships between numbers. It is important to know that thirteen is bigger or more than twelve and smaller or less than fourteen. Whether or not the child can read the numeral thirteen is still not important at this point in time.

It is not necessary or advisable to always count all the way to twenty. Count as far as your child is able, gently extending his or her range one or two numbers at a time. And once your child can count to twenty, vary the number of objects to be counted to ensure that counting is more than simply reciting words without deeper comprehension.

Chances are that you and your child will not count to twenty together as often as you counted to ten. Thank goodness! However, remember that counting is the most basic of math skills and the repetition is helping your child to develop a strong sense for numbers.

A good way to reinforce your child's learning is to ask him or her questions such as "what number comes after thirteen?" or "what number comes before eighteen?" It is easier to rhyme numbers off in order than to start in the middle and figure out what number comes either before or after.

Once your child has mastered counting to twenty, he or she will most likely be curious about what numbers, if any, come next. It certainly will not hurt to continue counting on past twenty. Your child will probably decide to count to some arbitrary number like twenty-seven and then stop. That, of course, is fine. Once your child is ready to move further ahead with counting, there are other strategies to employ other than just endless counting.

CHAPTER SEVEN

Trampoline Math

7

Few families actually have a trampoline around the house, but this activity can also be done on an old sofa, bed, or any other bouncy place. The idea, quite simply, is to have your child count while bouncing. If safety is a concern, hold your child's hands while they bounce.

Why bounce? There are a number of reasons. First and foremost, because it is fun and you want to make learning to count as much fun as possible. Secondly, because the rhythm of bouncing may help some children to more easily learn the rhythm of counting. Thirdly, because it is an excellent way for an active child to expend a little energy while learning. Finally, because it is a refreshing alternative to always using counting objects.

To begin with, of course, your child will only be learning to count by one's and jumping will be an enjoyable way for him or her to practice. Remember to use this activity when your child is learning to count backwards and when learning to count by two', five's, and ten's.

CHAPTER EIGHT

Subtracting from Ten by One's

Once your child can count backwards from ten with confidence, it is a good time to introduce the concept of subtraction. Counting backwards is, of course, a form of subtraction. Your child can learn the concept of subtraction by making it more explicit as you count backwards.

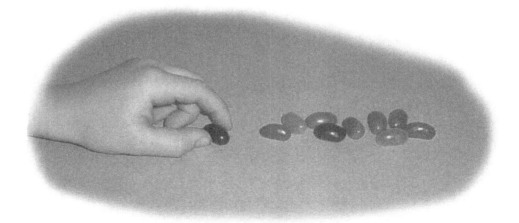

When counting backwards from ten start adding words such as "take away one equals" or "subtract one more makes" and allow your child to supply the answer. As suggested above, even though you are essentially just counting backwards, the new words may cause your child to hesitate before answering. By saying those words you are interrupting the now easy rhythm of counting backwards and making your child have to think about the answer.

MARSHMALLOW MATH

   Continue to use objects, such as macaroni noodles, pennies, or jellybeans to illustrate the point. Even if your child can count backwards from ten easily, seeing the objects being removed helps to reinforce the concept of subtraction. The decreasing number of objects will clearly demonstrate the concept of decreasing quantity. Encourage your child to count out all of the remaining objects before giving an answer. This will keep him or her from simply counting backwards from memory. For the same reason, vary the number of objects you set out.

   Once your child has given the correct answer, with or without your assistance, confirm and reinforce the answer by quickly repeating the exercise. State out loud what you are demonstrating. For example, "seven take away one equals six."

CHAPTER NINE

Orderly Ordinals

9

Ordinal numbers simply indicate the "order" of people, things, and events in a sequence. When we use the words "first", "second", "third", "fourth", "fifth", "sixth", "seventh", "eighth", "ninth", "tenth", etc. we are using ordinal numbers. Cardinal numbers, on the other hand, are simply the numbers that we use to count with. When your child tells you that he or she has six marbles, he or she is using cardinal numbers.

Very likely, you use ordinal numbers when talking to your child all of the time. To teach ordinal numbers to your child, simply become more aware of your own use of ordinals and place a greater emphasis on them when talking to your child. Find opportunities to use the words first, second, third, etc. "Your sister will have her bath first and you will go second." "We will read this book first, this book second, and this book third." Perhaps, have your children form short queues to do things such as brush their teeth and wash their hands. Indicate who

MARSHMALLOW MATH

is going to go first, second, and third. At the risk of starting a quarrel, when starting a new activity ask your children who wants to go first and who wants to go second. Similarly, when planning the day's activities ask your children what would they like to do first, what would they like to do second, etc.

If you only have one or two children, you may want to improvise using dolls or teddy bears. Dolls and teddy bears make good teaching aids for all but the largest of families when introducing ordinal numbers such as seventh and eighth.

CHAPTER TEN

Adding to Ten Again 10

Been there, done that? It is time to mix things up a bit more by adding together any two numbers where the sum total is not more than ten. For example, you might ask your child to add "six plus three" or "six plus four", but not "six plus five". Ten makes a good cutoff point because in the future, once your child begins to add together larger numbers, it will be particularly helpful to recognize all of the combinations that make ten.

At this stage, adding is still just a variation on counting. Your child will still need objects to count. Place various numbers of macaroni noodles, or other small counting objects, into two small lines or clusters. Ask your child to count out how many there are in the first cluster and then to count out how many there are in

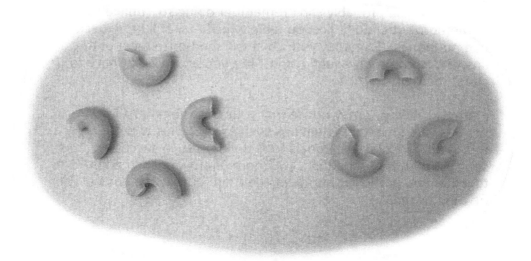

the second cluster. Then ask him or her how many there are all together if the two clusters are put together. Move the two clusters closer together, but avoid combining the two clusters into one big pile. Have your child count all of the objects in both clusters. Once your child has given the correct answer, reinforce their achievement by pointing at the two clusters of objects and repeating, for example, "four plus three equals seven".

When counting out the total number of objects in both clusters, your child will likely start counting again beginning with one. For example, if your child is adding four plus five and has already counted out both the cluster of four and the cluster of five, he or she will not start counting at four or at five, but will start again from the very beginning at one. This may be frustrating for you, but is a normal part of child development. It will likely be some time before your child will start counting from either of the numbers represented by the two clusters of objects. Eventually he or she will understand that they do not have to start counting right from the beginning, but it will take time.

Do not expect your child to quickly memorize all of the answers. You may "just know" that "four plus three equals seven", but for a period of time your child will have to count out the answer each time he or she attempts that problem. You "just know" the answer because you memorized it a long time ago. So long ago, in fact, that you probably can't remember doing so, (unlike trying to memorize the multiplication table which may still bring back bad memories). Someday your child will memorize most of the answers to common equations just like you and will not have to count them out.

After a while you may notice that your child does "just know" the answers to certain equations such as "two plus two equals four" and "five plus five equals ten". Build on such successes. It is simply a matter of repetition. The more often your child adds together two particular numbers, the more quickly the answer will be memorized.

For quick reference, all of the simple addition problems that result in a sum total of not more than ten are listed below. Be sure to mix them up.

One plus one	Two plus seven	Five plus one
One plus two	Two plus eight	Five plus two
One plus three	Three plus one	Five plus three
One plus four	Three plus two	Five plus four
One plus five	Three plus three	Five plus five
One plus six	Three plus four	Six plus one
One plus seven	Three plus five	Six plus two
One plus eight	Three plus six	Six plus three
One plus nine	Three plus seven	Six plus four
Two plus one	Four plus one	Seven plus one
Two plus two	Four plus two	Seven plus two
Two plus three	Four plus three	Seven plus three
Two plus four	Four plus four	Eight plus one
Two plus five	Four plus five	Eight plus two
Two plus six	Four plus six	Nine plus one

CHAPTER ELEVEN

Don't Forget Zero

11

It is easy to forget that zero is a number too when teaching young children to count, add, and subtract. Fortunately, children love the number zero and seem to be fascinated with the idea that we give a number for nothingness.

You can have lots of fun with zero. Ask your child silly questions such as, "How many monkeys are sitting at our table," or, "How many watermelons do you have in your pocket?"

When counting start from zero. When counting backwards finish with zero. Remember to both add zero to other numbers and to subtract zero from other numbers.

CHAPTER TWELVE

Finding Tens

12

As mentioned above, with our base ten number system the number ten is of great significance. When adding several numbers together, it is very helpful to be readily able to identify combinations of numbers that add up to ten. As well, in the future, your child will find doing more complex problems easier if he or she can quickly spot combinations of ten.

Let's say that you want to add up the following numbers, "6 plus 7 plus 2 plus 3 plus 8 plus 4". Adding the numbers in the

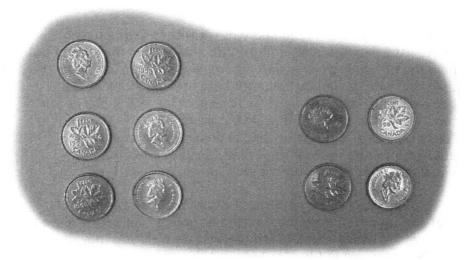

order written takes considerable effort. However, if you note that "6 plus 4 equals 10", that "7 plus 3 equals 10", and that "2 plus 8 equals 10", it is easier to add all of the numbers together.

MARSHMALLOW MATH

By placing special emphasis on learning the following equations, your child will memorize them sooner. The following are all of the simple number combinations that add up to ten.

Zero plus ten
One plus nine
Two plus eight
Three plus seven
Four plus six
Five plus five

Six plus four
Seven plus three
Eight plus two
Nine plus one
Ten plus zero

A good activity is to count out ten pennies or other counting objects in a straight line. Then illustrate the above combinations by separating the objects apart. Move one object over at a time and then count how many have been moved and how many remain. Recite the addition equation out loud. For example, "one plus nine equals ten." Gradually, the number of objects on one side decreases as the number of objects on the other side increases. Repeat the process in the other direction. Emphasize how the total always adds up to ten.

Once your child can read and write numbers, print out the above list using numerals and explore it with your child. Point out how the numbers on one side get smaller while the numbers on the other side get larger. Point out how the equations at the bottom of the list are simply the reverse of the equations at the top. That is, for example, that "two plus eight" is the same as "eight plus two". Once your child is doing mental math, start asking him or her questions such as "six plus what number equals ten?" or "what number plus six more equals ten?"

For more practice memorizing the combinations of numbers that add up to ten, play the Covered Counting Game, described in Chapter 29, using a total of ten objects. As well, you may use dice as described in Chapter 13.

CHAPTER THIRTEEN

Lucky Dice

13

Becoming proficient at adding, like any other skill, takes a lot of practice. As a good parent, you will want to give your child lots of opportunity to practice adding numbers together. However, as a busy parent, you do not have lots of time to spend preparing lessons or making up exercises. Using a pair of dice is a great way to give your child great practice adding. It is quick and fun. Your child can easily confirm and reinforce the answers to the more difficult addition problems by counting up the dots. Further, this game can grow in difficulty as your child's skill increases.

Begin by having your child become familiar with a single die. What numbers do the dots represent? How many sides are there on a single die? How many numbers are there on a single die? Encourage your child to become familiar with each of the dot patterns so that he or she can quickly identify each number without counting out the dots.

Once your child can identify each number pattern without counting the dots, throw a pair of dice. First, ask your child to identify each of the two numbers that turn up. Next, ask him or her to add the two numbers together. If your child has trouble doing so, suggest that he or she count the dots. Once your child arrives at the correct answer, reinforce the answer by repeating the problem in words. For example, "six plus three equals nine." Then roll the dice again. Each roll will present your child with another addition problem to work on. After a short while, the same two numbers will reappear. Such repetition is a great way for your child to memorize the answers to common addition equations.

The best thing about using dice to practice addition is that the difficulty level can easily be adjusted to match your child's skill level. At the beginning your child may have to physically count out each answer. After a while, your child will be able to count out the answer mentally. Later, he or she will begin to memorize answers and will be able to shout them out. Obviously, the game can then move along more quickly. Once your child can count any two numbers with confidence, add a third die. Suddenly, the challenge becomes much greater. In time, add a fourth and a fifth die. By then, you may well find the game as challenging for you as for your child. Adding with dice will become a wonderful exercise in mental mathematics for you and your child.

Playing with dice is also a good way to give your child practice identifying combinations of numbers that add up to ten. Roll two or three pairs of dice all at once and then ask your child to find all of the combinations that add up to ten. With most rolls, one or more combinations that add up to ten will appear. Older children may find combinations that add up to ten using three or more numbers.

Consider making a customized pair of dice. Take an ordinary pair of dice and glue on the numbers 5, 6, 7, 8, 9, and 10. While there will not be any dots to count, it will expand the number of combinations that may be rolled. Either roll a pair of customized dice together or roll a customized die with a regular die.

Adding Numbers from Zero to Ten

14

Earlier, it was suggested that a natural step in the ongoing development of addition skills is to add together any two numbers that result in a sum no greater than ten. Obviously, at some point you and your child will want to progress beyond that. Another natural phase or level in competency is to add together any two numbers from zero to ten. For example, you might ask your child to add together the numbers "seven plus eight" or "ten plus nine", but not "twelve plus two".

To some extent such a cut off is simply arbitrary, but there are good reasons for doing so. It gives your child lots of experience

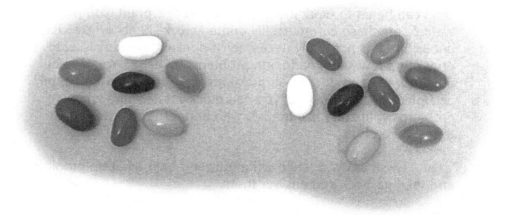

adding together all of the single digit numbers as well as the number ten. Such numbers form the core of all addition problems and deserve special attention. The equation "twelve plus two" given above, for example, can be broken down into "ten plus two plus two". A child who learns to add together all of the numbers from one to ten with confidence will be able to add larger numbers together with ease.

At first, it will still be necessary to use counting objects such as jellybeans. Over time, however, your child will begin to depend less and less upon physically counting out answers. Counting out up to twenty jellybeans will become tedious. Nevertheless, your child will still be counting out most answers in his or her head until he or she memorizes the answers or learns to add abstractly. Eventually, your child should memorize all of these basic addition facts.

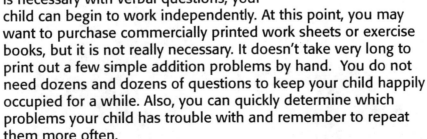

As well, once your child learns to read and write numerals well you may begin printing out addition questions, and later subtraction questions, for him or her to work on. Rather than working through each question together, which is necessary with verbal questions, your child can begin to work independently. At this point, you may want to purchase commercially printed work sheets or exercise books, but it is not really necessary. It doesn't take very long to print out a few simple addition problems by hand. You do not need dozens and dozens of questions to keep your child happily occupied for a while. Also, you can quickly determine which problems your child has trouble with and remember to repeat them more often.

For a twist, ask your child write out their own addition problems and then go back and answer them.

There are a number of ways in which you may quickly generate questions for your child. You may simply pick randomly from the list below. Alternatively, remove the face cards and jokers from a deck of playing cards. This will leave cards two through ten and aces to represent one's. Shuffle the remaining deck and deal out cards to add together. Another idea is to purchase or make a number spinner. You may use a spinner taken from a children's game. If it does not already have numbers from zero to ten on it, simply glue numbers on. A third idea is to make a customized pair of dice. Glue numbers 5, 6, 7, 8, 9, and 10 onto a regular die. To generate questions, either roll a pair of customized dice together or roll a customized die with a regular die.

For quick reference, all of the simple addition problems using numbers from zero to ten are listed below. Choose randomly from among the following questions.

MARSHMALLOW MATH

Ten plus zero	Eight plus six
Ten plus one	Eight plus seven
Ten plus two	Eight plus eight
Ten plus three	Eight plus nine
Ten plus four	Eight plus ten
Ten plus five	Seven plus zero
Ten plus six	Seven plus one
Ten plus seven	Seven plus two
Ten plus eight	Seven plus three
Ten plus nine	Seven plus four
Ten plus ten	Seven plus five
Nine plus zero	Seven plus six
Nine plus one	Seven plus seven
Nine plus two	Seven plus eight
Nine plus three	Seven plus nine
Nine plus four	Seven plus ten
Nine plus five	Six plus zero
Nine plus six	Six plus one
Nine plus seven	Six plus two
Nine plus eight	Six plus three
Nine plus nine	Six plus four
Nine plus ten	Six plus five
Eight plus zero	Six plus six
Eight plus one	Six plus seven
Eight plus two	Six plus eight
Eight plus three	Six plus nine
Eight plus four	Six plus ten
Eight plus five	Five plus zero

Five plus one

Five plus two

Five plus three

Five plus four

Five plus five

Five plus six

Five plus seven

Five plus eight

Five plus nine

Five plus ten

Four plus zero

Four plus one

Four plus two

Four plus three

Four plus four

Four plus five

Four plus six

Four plus seven

Four plus eight

Four plus nine

Four plus ten

Three plus zero

Three plus one

Three plus two

Three plus three

Three plus four

Three plus five

Three plus six

Three plus seven

Three plus eight

Three plus nine

Three plus ten

Two plus zero

Two plus one

Two plus two

Two plus three

Two plus four

Two plus five

Two plus six

Two plus seven

Two plus eight

Two plus nine

Two plus ten

One plus zero

One plus one

One plus two

One plus three

One plus four

One plus five

One plus six

One plus seven

One plus eight

One plus nine

One plus ten

Zero plus zero

CHAPTER FIFTEEN

Counting Backwards Again

15

Earlier, counting backwards from ten was suggested as a good activity for children learning number concepts. Children enjoy counting backwards from ten and will often do so on their own while playing. Once your child can count backwards from ten with confidence, suggest that he or she try counting backwards from a larger number such as fifteen or twenty.

Counting backwards is a good stepping-stone to subtracting because it helps your child learn the order of numbers in both directions. Children generally find adding easier than subtracting and this is probably because we are more familiar with the order of numbers getting larger. Subtracting is more difficult because we are less familiar with the order of numbers getting smaller. It doesn't come as naturally. While we often have many reasons to count forwards, we seldom have any reason to count backwards.

So do not be surprised if your child finds counting backwards from a larger number somewhat difficult at first. Counting forwards has been memorized like a poem in which each number suggests the next. When counting backwards, your child will have to work harder to remember the number order.

Once counting backwards by one from any number becomes relatively easy for your child, add to the challenge by suggesting that he or she count backwards by two's or three's.

Subtracting Numbers from Zero to Ten

16

At some point in time, you will want your child to begin subtracting numbers that are larger than one. As with addition, focus on subtraction questions that only use numbers from zero to ten. Subtraction, of course, is simply the reverse of addition. It is good to demonstrate that, for example, if "two plus three equals five", then "five subtract three equals two". This helps to illustrate the relationship between numbers and their wonderful consistency.

Continue to provide macaroni noodles or other objects for your child to count out. Start by having your child count out a given number of objects that is no greater than ten. Try to keep the objects in an orderly row. Then ask your child to "take away" or "subtract" a certain number of the objects. Have your child count out the number of objects to be subtracted and move them apart from the others. Next, ask your child tell you how many objects remain. Your child will have to count out the remaining objects

to determine the answer. That is to be expected for some time.
The concept of subtracting larger numbers is, nevertheless, being
learned.

Once your child has given the correct answer, with or without
your assistance, quickly re-enact the exercise. Move all of the
objects back into a neat row and then repeat the subtraction.
State out loud what you are demonstrating. For example, "eight
take away five leaves three." Then ask your child to count out a
different subtraction problem.

After subtracting the same numbers over and over many times,
your child will begin to memorize some of the answers. This
should be encouraged. It is a long process and repetition is the
key. However, there is no need to wait until your child has
memorized all of the answers before moving on to other skills.
Your child can be competent at subtracting without being able to
shout out the answer as soon as you state the question.

For quick reference, all of the simple subtraction questions
using numbers zero to ten are listed below. Choose from the
following questions randomly.

Ten minus zero	Nine minus zero
Ten minus one	Nine minus one
Ten minus two	Nine minus two
Ten minus three	Nine minus three
Ten minus four	Nine minus four
Ten minus five	Nine minus five
Ten minus six	Nine minus six
Ten minus seven	Nine minus seven
Ten minus eight	Nine minus eight
Ten minus nine	Nine minus nine
Ten minus ten	Eight minus zero

MARSHMALLOW MATH

Eight minus one
Eight minus two
Eight minus three
Eight minus four
Eight minus five
Eight minus six
Eight minus seven
Eight minus eight
Seven minus zero
Seven minus one
Seven minus two
Seven minus three
Seven minus four
Seven minus five
Seven minus six
Seven minus seven
Six minus zero
Six minus one
Six minus two
Six minus three
Six minus four
Six minus five
Six minus six
Five minus zero
Five minus one
Five minus two
Five minus three
Five minus four
Five minus five

Four minus zero
Four minus one
Four minus two
Four minus three
Four minus four
Three minus zero
Three minus one
Three minus two
Three minus three
Two minus zero
Two minus one
Two minus two
One minus zero
One minus one

CHAPTER SEVENTEEN

Lucky Two's and Terrible Three's

17

This is a simple adding and subtracting game to play with dice. Start by deciding on a number goal to reach for such as twenty. Each player takes a turn rolling a single die. The number rolled is added to that player's score. The first player to reach or exceed the number goal is the winner. If a player rolls a two, they get to roll a second time. If a player rolls a three, they have to subtract three from their score. You cannot have a score below zero.

Once your child is older, you may change the rules. Play with a pair of dice and choose larger number goals. Play lucky seven's and terrible ten's. If a player rolls any combination of seven, they get to roll a second time. If a player rolls any combination of ten, they have to subtract it from their score. As well, you may make it a rule that players must reach the target number exactly to win.

PART TWO

More Than Arithmetic

Part Two explores mathematical skills that do not involve arithmetic. It covers topics such as sorting and comparing, telling time, spatial awareness, pattern making, and elementary geometry. Part Two also discusses ways to help your child learn how to read and write numerals. The concepts and skills found in Part Two do not fit within the orderly progression of skills set out in the other chapters of this book and may be developed at the same time as other skills.

CHAPTER EIGHTEEN

Sorting and Comparing 18

Along with counting, you can have fun with very young children sorting and comparing various things around the house. Many activities of daily living provide good learning opportunities.

Classifying and naming the flora and fauna of the natural world according to basic traits or characteristics has long been a goal for scientists. Sorting, comparing, and classifying things will help your child to make sense of their world and learn to identify the differences and similarities between things. For example, although different breeds of dogs can look very distinctive, they all have certain traits or characteristics that make them dogs and not, for example, cats.

Encourage your child to help while you sort laundry to be washed; light coloured clothes in one basket and dark in the other. He or she will also have fun helping to sort clean laundry. Make it your child's job to find pairs of socks. Pull out the cutlery tray and encourage your child to sort and put away the forks, spoons and knives. After grocery shopping, mix-up your fruit and have your child sort out the apples from the oranges and other fruit. Toys may be sorted and classified into soft and hard toys, big and little toys, etc.

MARSHMALLOW MATH

Help your child to think about and discuss why they are sorting things the way they are. Assist your child in identifying and

describing the characteristics or attributes of the things being sorted and compared. What attributes do all of the things share? In what ways are similar things different? If you are sorting socks, for example, all socks share some attributes. All socks are soft and fit over a person's feet. As well, there is a hole at one end of each sock and no hole at the other end. Hopefully, socks all come in matching pairs. On the other hand, there are certain attributes that help to distinguish one pair of socks from others. Socks come in different sizes and in different colours.

Comparing different objects is also a good way to introduce your child to important mathematical concepts such as length, height, width, and weight. Guide your child's thinking about what makes one object "bigger" than another. Is it the length, the width, the height, or the weight of the object? Can one object be shorter, but heavier than another object? Can one object be taller, but narrower than another? Which would be bigger? What do we mean by bigger? Is "bigness" a useful concept?

There are many items around the house such as pots and pans, shoes, and books that are good for comparing. Have your child arrange pots and pans, shoes, or books according to their length, height, width, or weight. Such activities introduce the concepts of sequence and place order. If you want your child to have a lot of fun, and learn at the same time, let him or her arrange all of your canned goods on the floor. Ask your child organize the cans according to their size. Talk to your child about how cans may be different in weight, height, and circumference. You do not have to use big words like "circumference" with your child for him or her to gain insight into the concept. Ask him or her which cans are the biggest around.

CHAPTER NINETEEN

Concepts of Time

Our concept of time is, of course, very mathematical. We break time down into periods of various lengths such as years, months, days, minutes, and seconds. Understanding time requires an appreciation of the concepts of past and future. To understanding past and future we need to appreciate that events take place in an orderly continuum of time. A particular event, such as nap time, will take place after certain events, such as lunch, and before other events, such as play time.

Time, along with money and sex, is one of the big mysteries for children. Time is such a preoccupation for adults that it can be difficult to appreciate how abstract and complex a subject it is for young children. Young children live very much in the present, with little thought given to the past or the future. In part, this is because of their limited conceptual ability, but it also reflects their lack of experience.

Consider a child on his or her third birthday. The child will have no memories of his or her first birthday and only the vaguest, if any, memories of his or her second birthday. The language skills and conceptual framework required to have understood the significance of having a birthday would have been lacking. Therefore, the event would not have been deeply embedded into his or her memory. Without a good memory of his or her last birthday, or other similar yearly events such as Christmas or Easter, the child will not have a good concept about the time period of a year. Likewise, without memories of past yearly events, the child will have no means of anticipating that events such as birthdays, Christmas, and Easter happen each year. Without that sense of anticipation of coming events, the future must surely be a vague concept.

It is also worth remembering that a time period of a year is enormously long for children while being all too short for adults. The years seem to pass more quickly as we grow older because each year represents a smaller fraction of our total life span. For a three-year-old child, a year represents one third of his or her lifetime. Relatively speaking, it is equivalent to ten years for a thirty-year-old person or twenty years for a sixty-year-old person. In terms of learning, development, and character forming experiences, a year for a three-year-old is equivalent to a much longer period for adults.

Shorter time periods of a month or a week are no less nebulous for a young child. Dividing time into weeks and months is a rather arbitrary thing and there is no absolutely compelling reason why society has adopted the Gregorian calendar. Without regular significant events to mark the beginning and end of each week or month, it is difficult for children to conceptualize such time periods. For a child, one week flows into the next and one month flows into the next with little to mark the occasion. As young children do not go to school or to work or have other "commitments", there is little for them to anticipate in the future.

Time periods shorter than a day are also arbitrary and, therefore, difficult to conceptualize. Days have been divided into 24 hours for obscure astronomical reasons and hours have been divided into 60 minutes because, apparently, ancient Babylonians thought that 60 was an important number. Gaining a sense of how long an hour, half-hour, or a minute is takes "time" for children to grasp.

The time period that children can understand most easily is, of course, the day. Unlike months, weeks, and hours there is nothing arbitrary about the length of a day. The rising and setting of the sun mark a day. There are other daily events, such as breakfast, that children can both remember and anticipate. Your child's concept of time will naturally, therefore, revolve around these events. His or her day will be marked by breakfast time, lunchtime, naptime, dinnertime, and bedtime. Other events happen in relation to these daily events. Your child will start to remember that he or she went to the park after breakfast and anticipate that he or she will have a nap after lunchtime.

The passing of a day is marked, of course, by the coming of night. Going to sleep at night and waking to a new day is how many children begin to mark the passing of longer periods of time. As children begin to anticipate coming events such as going to the zoo or going to the beach, the number of days that they will have to wait will, somewhat ironically, be thought of as the number of times they will have to go to bed.

Another period of time that is not arbitrary and, therefore, easier for children to understand are the seasons. Your child will begin to notice the change in the seasons as winter becomes spring, spring changes into summer, summer turns into fall, and fall becomes winter again. Your child will remember past seasons and, with experience, will begin to anticipate the coming seasons. The four seasons provide a natural way to mark the length of a year.

Early in your child's life, there is little point in trying to teach him or her arbitrary concepts such as months or how to read a clock. Young children do not have the experience or conceptual framework to grasp such abstract ideas. However, there is much

that you can do to help your child to begin understanding time. Take advantage of the natural events that mark time to introduce concepts of time to your child. Talk to your child about daily events such as breakfast and lunch that mark the time of day. Tell your child what you will be doing together in the near future, after lunchtime or after dinnertime. Ask your child to remember what he or she did in the past, after breakfast time or before lunchtime. Explain the concepts of "morning", "afternoon", "evening", and "night". Introduce the concepts of "yesterday" and "tomorrow". Count down the number of days, or "big-sleeps", to special events such as a birthdays and Christmas. Note the change in seasons and talk about past and future seasons.

Your child will, of course, hear you talk about other periods of time. How many times have you replied to your child's many demands by saying "in a second" or "in a minute"? Eventually, your child will begin to question just how long a second or a minute really is. Likewise, your child will often hear you speak about hours. Relate such time periods to things that your child does on a daily basis. It takes about a second to take a big jump or to clap your hands twice. It takes about a minute to wash and dry your hands well or to brush your teeth. It takes about an hour to drive to drive to Grandma's house or to eat dinner.

Your children will also hear you refer to the days of the week. Once your child demonstrates curiosity about what day it is, you may want to teach him or her the days of the week. Anchor the name of the days down with regular events that your child can anticipate. "On Monday, mommy and daddy go to work." "On Sunday, we go to church." Help you child learn to recite the days of the week like a poem.

In Chapter 26, below, ways to help your older child learn to read the clock will be discussed.

CHAPTER TWENTY

Pattern Making

20

Children enjoy discovering and making patterns such as, blue - red – red – blue – red – red. Pattern recognition is an important skill to develop as it helps us to make sense of information that might otherwise appear to be nothing more than a chaotic mess of data. It is fundamental to learning to read both words and music, creating art, analyzing information, solving problems, and other skills. Working with patterns helps children to develop important thinking skills. Pattern making activities will help your child learn to see the individual parts of a whole, (that is to "analyze"), and to see how small parts may be put together to create a whole, (that is to "synthesize"). Identifying and creating patterns will help your child to see the relationships between and among things.

Once you begin looking, you will find many objects and materials around the house that are suitable for making patterns. For example, your child may use cutlery to place knives, forks, and spoons in various orders. Alternatively, use pennies, nickels, dimes, and quarters. A deck of cards can be used to create a variety of different patterns. Water paints and crayons can be used to make colourful patterns. The musically inclined can clap a rhythmic pattern. There are multitudes of different beads available in arts and crafts stores that are ideal for making patterns. Your child will enjoy making colourful patterns for bracelets and necklaces. This is also a wonderful activity for developing fine motor skills.

Start off by making simple patterns yourself and showing them to your child. Encourage your child to continue your patterns. Once your child learns to recognize and continue your patterns, invite him or her to start making their own patterns. Your child will be thrilled if you attempt to continue a pattern that he or she has made. Talk to your child about the patterns that they are discovering or making. Can your child describe the rules that the pattern follows?

Look for patterns around the house. You will find interesting patterns on floor and wall tiles, drapery material, wallpaper, pottery, clothing, etc. Talk about how patterns are used to make things look more attractive. Encourage your child to draw or model patterns that they find.

Much to the chagrin of parents, children love to play with their food. Take advantage of this inclination by encouraging your child to make patterns with food. He or she might help to arrange a cracker and cheese plate or a vegetable and dip plate. While eating, some children like to eat their food in a particular order. Have your child describe the pattern in which he or she is eating their food.

With older children, explore number patterns. You can make simple number patterns using counting objects, such as pennies, by grouping different numbers of pennies in a particular order

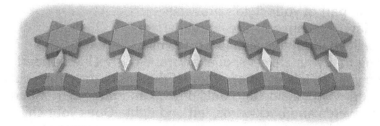

and then repeating the same groupings. For example, two pennies – four pennies – two pennies – four pennies. Some number patterns do not repeat. For example, the patterns made by adding by two's or adding by three's. You may also explore number patterns with a hundred number board. The hundred number board is discussed in Chapter 33.

CHAPTER TWENTY-ONE

Spatial Awareness

21

Spatial sense may be thought of as awareness of our physical environment and of the objects within that environment. Developing a spatial sense is important for living, working, and playing in our three-dimensional world. While the study of geometry will help older students to better understand two-dimensional and three-dimensional figures and shapes, young children benefit from exploring basic spatial concepts such as up and down, inside and outside, top and bottom, front and back, forwards and backwards, left and right.

While you might consider such words to simply be part of your child's growing vocabulary, the concepts they describe are essentially mathematical. It is worthwhile ensuring that your child's spatial awareness is well grounded. Explore your child's understanding of the positional words such as up and down, inside and outside, top and bottom, front and back, forwards and backwards, left and right. Does your child know when he or she is going upstairs and downstairs? Can your child identify the inside and outside of a box? Can your child identify the top and bottom of a table, the front and back of a book? Does he or she know what it means to move forwards and backwards? Can he or she identify his or her left and right hands and follow directions to move to the left or to the right.

Can he or she describe the position of one object, such as a toy, in relation to another object such as a table? For example, that "the toy is on, below, or beside the table".

Play a "treasure hunt" game with your child. Begin by hiding a "treasure", such as a small toy or perhaps a treat, somewhere in a room. Then ask your child start the hunt by standing in the middle of the room. Guide your child to the treasure by using the spatial directions, forward, backwards, left, right, up and down. Once your child gets the idea of the game, change roles. Let your child hide a treasure and then guide you to it.

Children, of course, enjoy making things out of construction paper and other materials. They also enjoy building things using toy construction sets such as "Lego" and "Meccano". Such activities are excellent for helping children develop spatial awareness and become competent at manipulating objects in two and three-dimensions. It also develops your child's fine motor and perceptual skills. Give your child lots of opportunity to do arts and crafts and to build things. Encourage your child to express their creativity.

CHAPTER TWENTY-TWO

Special Shapes

22

There is much more to math, of course, than adding, subtracting, multiplying and dividing. Many real life math problems will involve shapes and such geometry concepts as angles, circumference, perimeter, and area. Young children enjoy learning to identify different two-dimensional shapes such as circles, ovals, triangles, squares, and rectangles. Older children will learn to identify more complex two-dimensional shapes such as pentagons and three-dimensional shapes such as cubes, cones, cylinders, and spheres.

You may want to make or purchase a set of two-dimensional shapes made out of wood or plastic for your child to explore and play with. However, you can easily explore shapes just using a pencil and paper and a ruler. Use a ruler and a right-angled triangle to draw basic geometrical shapes on paper or cardboard to cut out. For circles, find a round cup or glass to trace. Help your child learn to identify the shapes.

Once your child can identify a shape, such as a rectangle, start looking around your home to find more examples of it. Look, for example, at windows, doors, cupboards, and tables. You may be surprised at how many squares, rectangles and other shapes there are to discover together once you start looking. Your kitchen cupboards will most certainly contain many circular shapes and, hopefully, one or more ovals. Fold paper napkins to make triangles.

Once your child can readily identify and find basic shapes, start comparing the basic characteristics of different two-dimensional shapes. Introduce the concepts of sides and points. Ask your child to determine how many sides and points each shape has. Take a careful look at how long different sides are. Are they equal in length or different? Explore how different shapes relate to each other. Ask your child to draw a square inside a rectangle, a triangle inside a square, a triangle inside a circle, and smaller triangles inside larger triangles. Such explorations will develop your child's intuitive sense of geometry.

You should also be able to find many examples of three-dimensional shapes around the house as well. Soup cans and toilet paper rolls are cylindrical in shape. Birthday hats and funnels are both cone shaped. You will find many spheres, of course, masquerading as balls. You may have to modify a box to make a perfect cube shape. Help your child to relate three-dimensional shapes to their two-dimensional cousins, such as squares and cubes, circles and spheres, triangles and cones. Compare different three-dimensional shapes and describe their

similarities and differences. Count how many sides and points there are in different three-dimensional shapes.

Blank number tiles, (that is upside down number tiles), are useful for making right-angled shapes. See Chapter 24. Help your child to make squares and rectangles out of blank tiles. Explain the difference between squares and rectangles. Ask your child to count how many tiles wide and long each shape is. Also, ask your child to count how many number tiles there are in each shape altogether.

Demonstrate to your child the concept of symmetry. Show him or her how both sides of a symmetrical object, such as bottle, look the same. Hold objects, such as toys, against a mirror to show how symmetrical images may be made. Look around the house for examples of symmetrical objects and designs. Look also for examples of symmetry in nature. Many trees and flowers grow symmetrically. You may make a symmetrical design by cutting or painting a piece of paper that is folded in half.

Tangrams

A tangram is an ancient Chinese puzzle consisting of seven shapes cut out of a square. The shapes consist of two large triangles, three smaller triangles, one square, and one parallelogram. An amazing number of different shapes and pictures can be made out of the seven shapes. You can purchase plastic tangrams at school supply and educational toy stores. Alternatively, you can make your own tangram out of construction paper or cardboard. You may photocopy the template found in the appendix of this book.

Tangrams will provide hours of fun for your child and help him or her to develop their perceptual abilities and understanding of elementary geometry. There are entire books and even games that are dedicated to tangram patterns. The following are only a few examples of what you and your child may do with tangram shapes.

What children like to do most with tangram shapes is to make up stylistic looking pictures. If you have an artistic flair, you can make up pictures of things like cats, foxes, boats, trains, rocket ships, and the like. Show it to your child and then take it apart and have your child try and make the same picture. As well, encourage your child to make up his or her own pictures. If you lack artistic flair, and even if you don't, you may also help your child learn how to create formal geometrical shapes such as squares, triangles and rectangles.

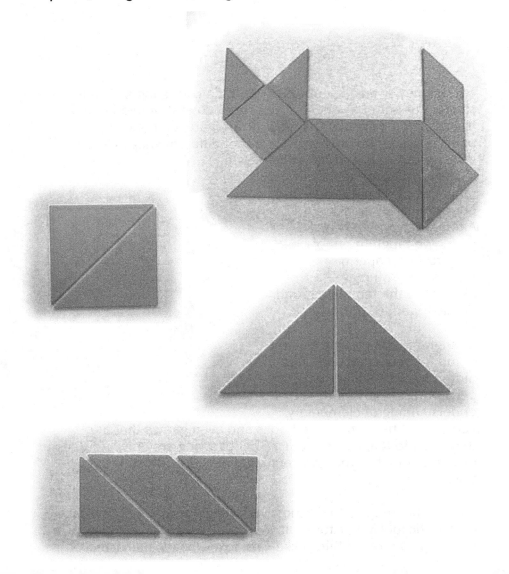

Reading and Writing Numerals

24

There are several good indications that it is appropriate to begin teaching your child to read and write numbers. One indication is that your child has developed a good understanding of number concepts. He or she counts well, understands the concept of quantity that numbers represent, and has started to manipulate numbers by adding and subtracting. A second indication is that your child has developed his or her fine motor control and has gained some skill in printing the alphabet. Perhaps the best indication, however, is simply that your child has begun asking you what numbers look like and how to print them.

Your objective now is to help your child learn to recognize the

symbols, or numerals, that we use to represent numbers. Your child knows what "eight" of something means, knows what number comes before and after "eight", and may even be able to add and subtract using the number "eight"; and yet may have no idea about what the printed numeral "8" looks like. However, learning to read and print numerals will likely come quickly once your child is ready.

There are many ways to help your child learn to read and write numerals. A good way to start is to simply print, as neatly as you can, numerals zero to ten and read them out loud to your child. After a while start encouraging your child to read the numbers back to you as you point at each number. Once your child is able to print, ask your child to first trace and then to copy numbers that you have neatly printed out. Print oversized numbers down the left side of a lined piece of paper and have your child copy out the numbers along the same lines. If your handwriting is not worthy of emulation, there are commercial materials available for your child to trace and copy. It is difficult to print numbers and it will take some time before your child will do so skillfully. Do not be alarmed if numbers are sometimes reversed, unless this persists for several years.

Occasionally, take the time to reinforce the relationship between printed numerals and the concepts of number and quantity. One idea is to print out numerals zero to ten and have your child line up the correct number of jellybeans, or other counting objects, below each of the printed numerals.

There are many others ways to give your child practice reading and printing numbers. An exciting way to encourage your child to read numbers is to guide him or her on a number safari and go "hunting" for numbers together. Once you and your child begin looking, you will find printed numbers everywhere. Your child will gain a greater appreciation for the importance of numbers in our lives.

Find page numbers in books. Look for numbers on the telephone, computer keyboard, TV remote control and, of course, calculator. Read the numbers on clocks, both traditional analog clocks and digital clocks. Introduce your child to the calendar and, perhaps, begin keeping track of the date together. Have your child look for house numbers as you drive along a street. Take your child grocery shopping. All of the price signs form a virtual jackpot of numbers. Read license plate numbers in parking lots. Look for numbers on signs while driving along the highway. Numbers, hopefully small ones, can also be found in the endless flow of bills arriving in the mail.

Here are a couple more ideas to consider. Have your child print out numbers in the sand, in a tray of flour, or while finger painting. Your child will also enjoy making numbers out of play dough or, even better, while baking cookies.

There are several teaching aids that you can use to help your child learn to read numbers. Number puzzles are excellent for helping young children learn to read numbers and to place them in the correct order. As well, there are innumerable children's picture and colouring books that are dedicated to helping children recognize numbers.

Another old standby is the dot-to-dot colouring book. Have you ever wondered about what the attraction of a dot-to-dot colouring book is for children? It certainly isn't the quality of the pictures created by connecting all of the dots! Dot-to-dot pictures usually look rather jagged and unsightly, even for children, compared to regular colouring books. Children enjoy dot-to-dot colouring books because they help to satisfy their intrinsic desire to learn. Connecting all of the dots in a dot-to-dot picture is an excellent way for your child to practice reading numerals.

Another teaching aid that can be used in creative ways to help your child learn to read printed numerals are number tiles. Number tiles from 1 to 100 may be purchased at school supply and educational toy stores. Alternatively, you may make your own out of cardboard. Number tiles are sometimes used on the blank side of a hundred number board. The hundred number board is one of the best teaching aids for teaching children about

MARSHMALLOW MATH

numbers and is discussed in Chapter 33 below.

A simple game to play using number tiles is number scramble. Take number tiles 1 through 10, (or higher depending on your child's skill level), and scramble them face up on the table. Then ask your child line up the number tiles in the correct order.

Another activity is to set out a particular number tile and then ask your child find the numbers that are one smaller and one larger than it. For example, if you set out number 4, have your child try and find numbers 3 and 5. Have your child continue adding on numbers in both directions.

When your child is more advanced, there are other variations on the scramble game to play. Scramble all 100 number tiles on the table face up. Have your child set out all of the even numbers in order or all of the odd numbers in order. Alternatively, have your

child set out number tiles counting by 3's, or 4's, or 5's, etc.

As well, you can use number tiles to quickly make up adding and subtracting questions. Further, blank number tiles are great for illustrating the difference between rectangles and squares.

CHAPTER TWENTY-FIVE

Fun and Games

25

There are many card and number games that provide a pleasant way of making numbers a part of your child's life. A few of them are discussed below. For other ideas, go to any good quality educational toy store.

Playing cards with your child is a enjoyable way to reinforce his or her number skills. Playing cards will help your child learn how to read numerals and remember number names. It also fosters an understanding of number order and an ability to recognize number patterns. Following the rules and developing strategy will help your child to develop logical thinking skills. Playing cards is also an excellent activity for enhancing your child's memory. And you thought that playing cards was just for fun!

MARSHMALLOW MATH

Go Fish, Snap, Pig, Donkey, Concentration, and Old Maid are just a few of the many card games that children enjoy playing. There are several books available that explain the rules to these and many other children's card games.

A classic counting game for young children is Snakes and Ladders. Playing Snakes and Ladders is a good way for young children to practice counting and learn to read numbers. With older children, playing Snakes and Ladders with two or more dice is a good way to incorporate addition practice into playtime. Frustration is another counting game, with a built-in dice roller, that young children enjoy playing. The object of the game is to be the first player to get all of your markers around the game board.

Dominos is another classic numbers game. It is a good game to help your child to learn numbers and recognize patterns. There are actually many different dominos games and, at higher levels of play, the game develops logical thinking and strategic planning skills. Bingo, of course, is also a good game for helping children learn their numbers.

For older children, Rummikub is a wonderful game for developing number and pattern recognition skills. The game is played with special coloured number tiles. Each player starts with twelve number tiles and the object of the game is to be the first player to lay down all of your numbers tiles. Players do so by laying down sets and runs of number tiles. Players can also add to sets and runs laid down by other players. A player who cannot lay down any tiles must pickup another tile from the remaining tiles.

Mancala is a fast-paced counting game from Africa. It involves moving forty-eight stones around a playing board that consists of twelve small bins and two larger "mancala" bins. The object of the game is to be the player with the most stones in his or her mancala bin. There is a considerable amount of counting and strategy involved in deciding which stones to move.

Other classic games that develop math and logical thinking skills include cribbage, backgammon, checkers, and chess.

CHAPTER TWENTY-SIX

Telling Time

26

Your child will let you know when he or she is ready to learn how to read the time on a watch or clock. He or she will have a good concept of such natural time periods as morning and afternoon, lunchtime, and dinnertime. He or she understands past and future, yesterday and tomorrow. He or she has learned to read and write numerals and will be very good at counting. In particular, it is helpful if your child is able to count by fives. Most important, your child has started asking what time it is and is no longer satisfied with being told that it is almost lunchtime. He or she will begin asking, "what time is it really?"

Once your child shows a serious interest in knowing how to tell time, a good way to begin is to simply relate everyday events to the formal "clock" time. Tell your child what time he or she goes to bed and what time he or she usually wakes up. Tell your child what time your family usually eats breakfast, lunch, and dinner. As well, begin answering your child's questions about time with more specific answers. "It is ten o'clock in the morning." "It is four-thirty in the afternoon." "Lunch will be in ten minutes". Your child will develop an understanding that time and numbers are related.

Even after your child begins to show an interest in what time it "really" is, it will be a big step to learn to read a clock. Today, of course, we have both digital and traditional analog clocks and your child will have to learn to read both kinds. Surprisingly, it does not seem to be that much easier for children to learn to tell time on digital clocks than on traditional clocks. Reading both types of clocks is difficult and will take a while to master. An advantage of using a traditional clock to teach your child about time is that it illustrates the cyclical nature of time. Night is followed by morning, morning is followed by afternoon, afternoon is followed by evening, and evening is followed by night.

An advantage of digital clocks, of course, is that the time is displayed in numerals that are easy to read. However, do not assume that this will make it easier for your child to truly comprehend.

Find a traditional clock, preferably one that is already broken so your child is free to abuse it, on which it is easy to rotate the hour and minute hands. Ideally, you should be able to move the hour and minute hands without touching them directly by turning a dial or pinwheel on the clock. Familiarize your child with the important parts of the clock: the minute hand, the hour hand, the minute markings, and the hour markings. Discuss how many hours there are in a day and how the day is broken into two twelve-hour periods. Explain about how hours are divided into sixty minutes. You may ignore seconds for now.

When telling time, begin by just reading out the hours of the day, keeping the minute hand pointing at the twelve. Relate the times shown on the clock to your child's daily routine. We eat breakfast at seven o'clock in the morning. We eat lunch at twelve o'clock noon. And so on. Once your child can tell you what hour the hands of a clock indicate, introduce the reading of minutes. Show your child how all of the minutes are marked off on the clock. Count all of the minutes in an hour.

MARSHMALLOW MATH

In reading traditional clocks, it is common to round off the time to the nearest five-minute point. It is usually close enough to say that it is ten minutes after the hour, when it may actually be nine or eleven minutes after the hour. One of the most difficult aspects of reading a traditional clock is to understand that the numbers marking the hours represent intervals of five minutes. Children are rather quick to conclude that it is three minutes after the hour when the minute hand is pointing at the three. That is why it is helpful for your child to be good at counting by fives in order to tell time. Practice telling the time in five minute intervals as you rotate the minute hand around the dial. For example, have your child tell you that it is five minutes after two o'clock, ten minutes after two o'clock, fifteen minutes after two o'clock, and so on. Once your child has some understanding of fractions, it is easy to illustrate concepts such as "a quarter after three" and "half past five".

One problem with learning to tell time with most digital clocks is that it is more difficult to quickly run through all of the different times of day. That is, it is difficult to quickly show one o'clock, two o'clock, three o'clock, etc. Also, it is difficult to introduce such ideas as rounding off time in five-minute intervals or in fractions of an hour. Digital clocks are so precise, even if they are actually wrong! However, it is important to be able to read and understand digital clocks. Show your child which numbers indicates the hour and which numbers represents the minutes. Encourage him or her to practice reading out the time.

Your child may have trouble reading a digital clock because while we read books from left to right, we often read digital clocks from right to left. For example, the time "5:12" is often read as being "12 minutes after 5".

To help your child learn to read of both kinds of clocks, try using both kinds of clock at once. Set both a digital and a traditional clock to the same settings. You might play a game whereby you set one clock to a certain time and have your child set the other clock to the same time.

CHAPTER TWENTY-SEVEN

Conservation: The Foundation of Logic

27

Children are in the process of developing the sophisticated thinking skills possessed by most adults. Young children, as every parent knows, lack logical, rational thinking skills. The thought processes of young children are marked by four adorable, but irrational, characteristics. Children's thinking has been described as being "perception bound". That is, children are overly influenced by the appearance of things. Children's thinking may be characterized as "centered". That is, children tend to focus on one detail and ignore other important facts. As well, children focus on the present state of things and not on how things got into that state. Further, young children are unable to reverse their thinking. That is, they are unable to go through a series of steps mentally and then reverse direction and retrace their mental steps back to the starting point.

The above thinking patterns are all illustrated by a young child's inability to solve what are called "conservation" problems. "Conservation" refers to the idea that certain physical attributes of an object remain the same, even if its outward appearance changes. The most fundamental conservation concept is "conservation of number". An example of understanding the conservation of number is knowing that stretching out a line of pennies will not change the number of pennies in the line. Comprehending the conservation of number provides a foundation for logical thinking skills.

It will be fun to try the following activity with your child. Set out two identical rows of six pennies side by side with each penny being spaced evenly apart. Ask your child, "Are there the same number of pennies in each row?" He or she will likely reply, "Yes". Then, with your child watching, make one row longer by spacing the pennies wider apart and make one row shorter by spacing the pennies closer together. Now ask, "Are there the same number of pennies in each row?" A young child will very likely answer that there are more pennies in the longer row.

The above activity illustrates all four aspects of a young child's thinking process. The child is "perception bound", being trick by appearances into believing that the longer row must have more pennies in it. The child's thinking is "centered", concentrating too much on one detail, the length of a row, and ignoring other important factors such as counting out just how many pennies there are in the row. The child focuses on the present state that the pennies are in and ignores how the pennies got into that new state. Lastly, the child is unable to reverse his or her thinking and imagine the pennies being pushed closer together again.

Until a child masters the concept of conservation of number, his or her understanding of numbers will be fuzzy at best. Conservation of number is a foundation skill. Your child will come to understand that the amount represented by a number does not change even if circumstances change. He or she will come to understand that six cows remain six cows whether they are standing close together or standing far apart. He or she will come to appreciate that six chickens is the same number as six cows, even if cows are bigger. He or she will come to realize that six chickens will always be more than five cows, even if cows are bigger.

Mastering the concept of conservation of number and other similar concepts requires the development of logical thinking abilities. It is a long and gradual process. There has been a major debate between child development experts as to whether or not instruction and experience can help a child develop logical thinking skills more quickly. Some believed that children must go through specific stages of development at their own pace and that until a child is cognitively ready, no amount of instruction or experience will help that child to grasp such concepts. However, more recent research suggests that instruction and experience can help children to develop logical thinking skills. That is, a child's cognitive abilities will develop in response to instruction and experience.

However, you cannot simply "teach" your child conservation of number and expect him or her to somehow "memorize" the concept. Your child will have to independently acquire the logical thinking skills that are needed to understand the concept. It is important to avoid trying to force the learning process. Most children do not achieve number conservation until between the ages of six and seven. With exposure to appropriate activities, your child will have the experiences that are necessary to gain insight into how numbers work. With experience, his or her thinking skills will mature. If you are counting and exploring other number concepts with your child, you are already providing your child with an enriched learning environment in which his or her cognitive abilities will develop.

The more meaningful counting experiences are for your child, the more likely it is that he or she will deepen their understanding of the meaning of number. Count things that are important to your child. For example, count out two rows of treats for your child to choose between. Put more treats into one row, but make the row look shorter. Put fewer treats in the second row, but stretch the row out. Help your child to determine which row to choose. You may also help your child to develop logical thinking skills by asking him or her how they arrive at their answers and by gently suggesting new strategies or ways of thinking about numbers. If your child is being fooled by deceptive appearances, suggest that they count more carefully. If your child is focusing on one fact or detail and ignoring others, help them to become aware of other considerations.

As you now know, "conservation" in child development lingo has nothing to do with saving the environment. Conservation refers to the idea that certain physical attributes of an object remain the same, even though its outward appearance changes.

CONSERVATION: THE FOUNDATION OF LOGIC

In addition to conservation of number, young children will acquire insight into conservation of length, liquid, and mass. Older children will learn about conservation of area, weight, and volume.

You may try doing the following activities with your child to see if he or she has mastered the concepts of conservation of length, liquid, and mass.

Take two sticks that are the same length, such as popsicle sticks, and lay them side by side so that the ends of each stick are even. Ask your child, "Is each of these sticks just as long as the other?" Your child will likely say, "Yes". Then, with your child watching, move one stick forward so that it sticks out past the other stick. Now ask, "Are the two sticks the same length, or is one stick longer than the other?" A child who does not yet comprehend conservation of length will say that the one stick is now longer than the other is.

In a similar activity, show your child two identical glasses of water. Ask your child, "Is there the same amount of water in each glass?" Your child will likely reply, "Yes". Then, with your child watching, pour one glass of water into a shorter but wider glass. Now ask, "Now is there the same amount of water in each glass, or does one glass have more water?" A child who does not comprehend conservation of liquid will say that one glass, likely the taller one, now has more water in it than the other glass.

Similarly, show your child two identical balls of play dough or modeling clay. Ask your child, "Is there the same amount of clay in each ball?" Your child will likely answer, "Yes". Then, with your

MARSHMALLOW MATH

child watching, roll out one ball into a tube shape or short "snake". Now ask, "Does each piece still have the same amount of clay, or does one piece have more? A child who does not comprehend conservation of mass will say that one piece has more clay in it than the other piece does.

As with conservation of number, you cannot simply "teach" your child conservation of length, liquid, and mass. Your child has to acquire logical thinking skills to understand the concepts. However, many everyday activities will give your child the experience needed to gain a better understanding about how our physical world works. Measuring things will help your child learn to appreciate that moving objects around will not change their length. Pouring liquids from one container into a different shape and size of container and back again will help your child to learn conservation of liquid. Playing with play dough or modeling clay, as all children love to do, will help your child to comprehend conservation of mass. Talking to your child about what he or she is doing and seeing and what he or she is thinking may help him or her to develop more sophisticated thinking skills.

PART THREE

Taking It Higher

Part Three continues to build upon the concepts and skills developed in Part One. Part Three starts off by exploring the joys of mental math and then looks at ways to further develop counting, adding, and subtracting skills.

CHAPTER TWENTY-EIGHT

Mental Math

28

Once your child is adding and subtracting counting objects such as macaroni noodles, pennies, or jellybeans with confidence, it would be an opportune time to introduce "mental math". Occasionally forego setting out objects for your child to count and, instead, ask your child to figure out the answer in his or her head. You may help your child to visualize the problem by verbalizing a picture for him or her. For example, you may ask, "If you already have three bananas and I give you two more bananas, how many bananas would you have?"

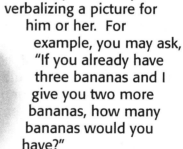

Do not be overly concerned about being original with each word problem. There is no need, for example, to always name a different object to count. You do not have to suggest, for example, apples for the first problem, oranges for the second, and coconuts for the third. You might always suggest that your child count apples in his or her mind and that would be fine. Your child's focus will be on the quantity of objects, not the objects themselves.

MARSHMALLOW MATH

From this point on, many of the activities can be done either mentally or by physically counting out objects. Alternate between both approaches. On the one hand, counting out objects remains a very important strategy for illustrating concepts, making abstract ideas more concrete, reinforcing learning, and building confidence. On the other hand, as your child gradually develops math skills and as you begin to use larger numbers, you will find counting out objects both more cumbersome and less necessary.

The biggest advantage to doing mental math, from a parent's point of view, is that you can really truly do it anywhere and anytime: while driving in the car, taking a bath, or walking the dog. Picture this; you are doing the dishes and your child is hanging around complaining, as children occasionally do, "I don't know what to do." Simply suggest that they count by threes to twenty-one or count backwards by two's from twenty, or any other challenge that is appropriate to their abilities. You will get a short break and your child will exercise his or her math skills. By the time your child completes the task he or she may well have found something else "to do" which, of course, is fine. Or they may ask for another math challenge, which demonstrates a positive attitude and should be encouraged. Mental math sessions can be as long or as short as the circumstances and your child's attention span determine.

Attempting to solve math problems without physically counting out objects will require a tremendous amount of concentration. The effort will be rewarded with a greater facility with numbers and the ability to focus on solving problems. While your child will still be counting out the answer in his or her head, he or

she will have to keep track of the numbers mentally. This is a difficult thing to do and represents an enormous step towards competency. After some time your child will begin doing calculations abstractly, without mentally counting out the answer.

CHAPTER TWENTY-NINE

Covered Counting 29

This activity is a good stepping-stone between having to physically count out each answer and being able to calculate answers mentally. It provides your child with physical objects to help him or her to visualize the question, but requires him or her to calculate the answer in their head.

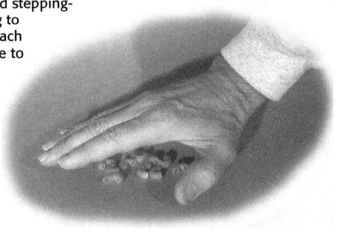

For practice adding, count out a number macaroni noodles with your child watching and then cover those noodles with one of your hands. Then count out a second number of noodles and add them to the noodles already hidden under your hand. Ask your child to tell you how many noodles there are under your hand altogether. For example, count out six noodles and cover them with one hand. Next, count out four more noodles and add them to those already covered by your hand. Once your child gives you the correct answer, ten in this example, remove your hand and have your child physically count out the total number of noodles. This will confirm and reinforce the correct answer and provides a sort of intrinsic reward for your child. He or she may become quite excited at being proven correct.

For practice subtracting, count out a number of noodles with your child watching and then cover all of those noodles with your hand. Then remove some of the noodles from under your hand. Ask your child to count how many noodles have been removed. Next, ask your child to calculate how many noodles remain hidden by your hand. For example, cover twelve noodles with your hand and then remove four noodles. Your child will have to use his or her mental math skills to calculate that eight noodles remain covered by your hand.

Until your child is skilled at doing calculations in his or her head, this can be a rather difficult activity. Your child will likely need to count out the answers, perhaps using his or her fingers, to begin with. Over time, with practice, he or she will become less dependent upon having to physically count out the answers.

CHAPTER THIRTY

Counting by Two's 30

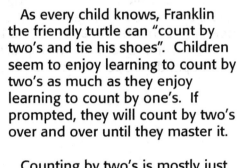

As every child knows, Franklin the friendly turtle can "count by two's and tie his shoes". Children seem to enjoy learning to count by two's as much as they enjoy learning to count by one's. If prompted, they will count by two's over and over until they master it.

Counting by two's is mostly just faster than counting by one's. As you begin to explore larger numbers, you will appreciate your child's ability to count by two's. Counting by two's is also a good foundation for understanding the concept of odd and even numbers in the future. You can highlight the fact that counting by two's is the same as adding by two's by saying words such as "plus two more" as you count out the pairs.

Counting by two's to ten is a natural first goal to aim for. There is no need to rush your child to count by two's past ten. Once counting by two's to ten becomes easy for your child, push ahead slowly. To begin with, your child may have trouble counting by two's. To make it easier, try suggesting that he or she say the missing odd numbers quietly under their breath and then say the next even number out loud. In time he or she will memorize the even numbers and counting by two's will become easier.

Remember to sometimes vary the number of objects that your child counts to and to sometimes count to an odd number. When counting to an odd number, such as seven or twenty-one, point out how one object does not have a partner.

Once your child can read numbers, print out a line of numbers and ask your child circle the even numbers as he or she counts by two's. With an older child, use a hundred board to count by two's all the way to 100. See Chapter 33.

CHAPTER THIRTY-ONE

Counting By Ten's

31

There are, of course, an endless number of numbers. It is natural to first teach your child how to count to ten and then to gradually extend their repertoire to eleven and twelve and all of the "teen's" to twenty. However, once your child can count to twenty with confidence learning to count higher accelerates quickly. This, of course, is because after "twenty" number names take on a predictable pattern. Children find knowing the pattern to number names to be a very empowering skill. Suddenly, they come to appreciate that they can count to very large numbers. The whole universe of numbers opens up to them.

Once your child can count to twenty with confidence, your next goal or target may jump all the way up to one hundred. Counting by one's to one hundred over and over with counting objects will be both impractical and unnecessary. The key to helping your child learn to count to one hundred is to teach him or her how to count by ten's.

While you will not always have to count out all of the numbers to one hundred using counting objects, it will still a good idea to use counting objects such as pennies to illustrate what counting by ten's looks like. Set out one pile of ten and then two piles of ten, and so on up to ten piles of ten. This should help make it clear to your child what it means to count by tens. It will also help him or her to appreciate how much one hundred really is. Discuss with your child how the number names are derived. There are two tens in twenty, three tens in thirty, etc. This will help him or her to remember the number names and to better appreciate of the size of each number.

Once your child knows what it means to count by ten's, have him or her do so over and over. It will be necessary for your child to repeat counting by ten's many times before he or she memorizes the number names. There are several strategies that you may employ. Have your child count by ten's to one hundred like a poem, "ten, twenty, thirty, forty, fifty, sixty, seventy, eighty, ninety, one hundred", to firmly lodge it in his or her memory. Have your child count to one hundred by ten's on his or her fingers. Once your child is familiar with the value of coins, and in particular that one dime is equivalent to ten pennies, counting out dimes is another good way of practicing counting by ten's. See Chapter 36. Using a hundred number board is also an excellent way to help your child learn how to count by ten's to one hundred. See Chapter 33.

Occasionally, ask your child to count backwards by ten from one hundred or start somewhere in the middle and count forwards or backwards by ten.

Once your child is counting by ten's, he or she will begin to pick up on the pattern of counting twenty-one, twenty-two, twenty-three, etc. all the way to one hundred. You may find, however, that your child will occasionally forget what number comes after, for example, forty-nine or seventy-nine and may need some prompting.

There are a number of activities that will help your child learn all of the numbers between twenty and one hundred. If your child is familiar with the value of coins, use dimes and pennies to illustrate two-digit numbers such as 23 and 47. Once your child

MARSHMALLOW MATH

understands what is expected, ask him or her to show you what various two-digit numbers look like using dimes and pennies. Play, "I'm thinking of a number". Tell your child, for example, that you are thinking of a number that has "three 10's and seven 1's". Ask your child tell you what number you are thinking of. Another idea is to select any two-digit number tile and ask your child to tell you what number it is. Then ask your child to tell you what number comes before and after that number.

Once your child understands the idea of counting by ten's to one hundred, even if he or she needs a little prompting now and again, there's no reason to not count higher. Show your child how the pattern of counting ten, twenty, thirty continues on past one hundred with one hundred and ten, one hundred and twenty, one hundred and thirty, etc. As well, why not introduce the concept of counting by hundred's to one thousand and then by thousands to??? Learning to count by hundred's and thousands may well be easier for your child than learning to count by one's and ten's. It is a matter of following a very straightforward pattern. Learning to count such large numbers will be exciting for your child and will give him or her confidence that they really are good with numbers.

32

Counting by Five's

While five is a smaller number than ten, it will be easier for your child to learn to count by fives after he or she has mastered counting by ten's. While counting by five's is not as important as learning to count by ten's, it is a useful skill that contributes to your child's overall ability in math.

Begin teaching your child to count by five's using counting objects such as jellybeans. Start by asking your child, how much does five plus five equal? He or she should be able to answer "ten" by memory. Then ask your child, "how much does ten plus five equal?" He or she will likely have to count out the answer "fifteen", which is ok. Continue along asking, "How much does fifteen plus five equal, how much does twenty plus five equal", and so on. Eventually, your child will begin to recognize the pattern and will be able to answer without counting it out.

MARSHMALLOW MATH

Once your child knows what it means to count by five's, there are a couple of ways to quickly practice the skill. One activity is to count dice. Take as many die as you have and turn up the number five on all of them. Then have your child use the dice to count by fives. Another activity is to count nickels. Collect twenty or more nickels together so that your child can count to one hundred or beyond with them. Using a hundred number board is another excellent way to reinforce your child's ability to count by five's to one hundred. See Chapter 33. Occasionally, ask your child to count by fives to one hundred like a poem to firmly lodge it in his or her memory.

CHAPTER THIRTY-THREE

The Hundred Number Board

33

One side of a hundred number board, or hundred board for short, sets out all of the numbers from one to one hundred in a 10- by-10 grid. The reverse side may set out a blank 10-by-10 grid on which number tiles may be placed. It is a wonderful tool for helping children learn counting and other number concepts because it helps children to visualize numbers and the patterns in numbers. A small hundred board may be found in the appendix of this book. Sturdy plastic hundred boards, and number tiles, are available at school supply and educational toy stores. Entire books have been dedicated to hundred board learning activities. This chapter will only discuss a few of the many learning activities that use a hundred board.

A hundred board is good teaching aid for helping your child learn to read numbers. In particular, it is an excellent tool for showing your child what larger numbers such as 43 and 78 look like. A good start is to simply point while counting out the numbers on the board. Show your child how all of the numbers with twenty in the their name start with the digit 2, that all of the numbers with thirty in their name start with the digit 3, and so on.

A hundred board is especially helpful in helping your child understand our base ten number system and to learn how to count by tens. Have your child count out how many numbers there are in the first row, the second row, the third row, etc. There are, of course, ten numbers in each row. Help your child to see that ten in the first row plus ten in the second row add up to twenty, that twenty plus ten in the third row make thirty, and so on. Practice counting by ten's to one hundred as you point at the numbers. Ask you child to tell you how many rows of ten there should be in, for example, forty or in seventy. Then have your child count the number of rows to confirm his or her answer.

You may highlight the number rows of ten by covering up all of the zero's in the 10's column so that it reads 1, 2, 3, etc. instead of 10, 20, 30, etc. Point to numbers out of order, for example 50, 20, 80, and ask your child to tell you which number you are pointing at.

Once your child can count by tens, use the hundred board to help him or her to learn all of the numbers between 1 and 100. Play a game of finding numbers. Pick any number between 1 and 100, for example 63, and ask your child to find it on the hundred board. Looking at all of those numbers can be rather confusing and your child may have difficulty finding a particular number. Help your child by suggesting he or she first find the 60's row and then count over to 63. Ask your child to tell you what number comes before and after a given number. For example, what number comes before 25 and what number comes after 25?

You may use the hundred board to practice "skip counting". Photocopy the hundred board found in the appendix of this book and then ask your child to circle or colour the numbers as he or she counts by two's, three's, four's, and five's.

The hundred board is useful for reinforcing adding and subtracting skills. For example, ask your child to find the number 2 and then add 6 more to it. Or, ask him or her to find the number 8 and subtract 6 from it. A hundred board is a particularly good method for showing children the "magic" of adding and subtracting by 10. Help your child to see the pattern that adding by ten makes on the hundred board. Pick a number such as 7 and ask your child to add 10 more making 17, and 10 more making 27, and so on. As well, you may try starting at a larger number, such as 54, and have your child subtract by 10's.

You may also play simple adding and subtracting games using a hundred board, a pair of dice, and a couple of markers to move about the board. Starting at number one, take turns rolling the dice and moving your marker according to the number you roll. See who gets to one hundred first. Alternatively, start at one hundred and work backwards using subtraction to get your marker back to one.

Adding and Subtracting Larger Numbers

34

With adding and subtracting, it was earlier suggested that you focus on questions that only use numbers from zero to ten. This is primarily because it is important for your child to memorize such basic addition and subtraction facts. All other addition and subtraction problems can be broken down into these fundamental components. Nevertheless, once your child is counting larger numbers with confidence, it is appropriate to challenge him or her with the addition and subtraction of larger numbers.

It is important to understand, however, that to begin with your child will simply be counting forwards and backwards as the case may be. For example, if you ask your child to add together "fifteen plus six", he or she will very likely count out the answer "sixteen, seventeen, eighteen, nineteen, twenty, twenty-one". He or she may either count out loud or in their head. If you give your child the question "eighteen minus five", he or she will likely count out the answer backwards, "seventeen, sixteen, fifteen, fourteen, thirteen".

While your child will memorize the answers to some common addition and subtraction problems that involve larger numbers, it is not reasonable to expect your child to memorize them all.

ADDING AND SUBTRACTING LARGER NUMBERS

There are too many of them. There are, in fact, an infinite number of them. Your goal will be to help your child learn to solve such questions abstractly, by working out the answer without counting it out. Your child will discover how to use the basic addition and subtraction facts that they do know to calculate the answers to new and more difficult questions. For example, your child may determine that the question "fifteen plus six" is the same as "fifteen plus five plus one". Likewise, your child may recognize that within the question "eighteen minus five" is an easier question "eight minus five".

When giving your child larger addition and subtraction questions to calculate, you may gently suggest ways in which they can break difficult questions down into easier parts. However, this will only take your child so far. He or she will also have to learn about the concept of "place value" in order to add and subtract very large numbers with skill. The concept of "place value" is discussed in Chapter 48.

CHAPTER THIRTY-FIVE

Fun with Fractions 35

Dividing things into fractions is a mathematical concept that your child may find easy to relate to because the concept is so often reinforced in everyday life. How many times as a parent have you; asked who wants half a banana, mixed in one-third of a cup of milk, or cut sandwiches into quarters?

When introducing your child to fractions, the colourful round, plastic or wooden fraction sets available at school supply and educational toy stores are highly recommended. Alternatively, it is a simple thing to make your own set of fractions. Use a saucer to draw four circles on coloured cardboard, preferably using four different colours. Leave one circle whole, cut one circle into two halves, cut one circle into three equal thirds, (at least as close as you can), and cut one circle into four equal quarters.

The first concept for your child to comprehend is; "What is a whole?" You can use the whole circle as an example of what is a whole. However, there are many better examples, and most of them can be found in your kitchen. Take an apple, for example, and explain to your child that it is a whole apple. Then take a big bite out of the apple and ask your child if it is still a whole apple. Help your child to understand that it is no longer a whole apple because part of it is missing. Only part of the apple remains. Another easy example is a glass of water. Fill a glass full to the top with water. Get your child to agree that it is a whole cup of water. Then pour out or, if you prefer, drink some of the water. Ask your child if it is still a whole glass of water. It cannot be a whole glass of water because some of it is gone.

Look around the kitchen for other examples of whole things. Can your child find a whole bag of chips or cookies, a whole carton of milk, a whole loaf of bread? Look around the kitchen for examples of things that are no longer whole. Can your child find a half-empty jar of jam, a half-empty box of cereal, or a half-empty bottle of ketchup? Keep your eyes open for things that are whole but won't be for long.

As well, ensure that your child knows what a part of something means. Again, the kitchen affords endless examples. Cut off a piece of a carrot or a banana, for example, and show it to your child. Ask him or her if it is the whole carrot. Hopefully, he or she will say "no". You should agree and explain that it is only a part of the carrot. When offering your child a cookie, ask whether he or she would like part of a cookie or a whole cookie. His or her answer will confirm how well they grasp the concept.

MARSHMALLOW MATH

Once your child has a good comprehension of what a whole of something is and what a part of something is, you can begin to introduce the idea of fractions. It will likely take a while before your child can independently identify and name the common fractions. The names themselves, half's, third's, and quarters are somewhat strange and unusual. Expect to go over them numerous times with your child.

Begin by setting out the four fraction circles in front of your child. Ask him or her to identify which circle is a whole one and which circles are divided into parts. Then focus on the circle that is divided into two halves. Ask your child to count how many parts there are. Obviously, there are two parts. Have your child compare the size of the two parts. Emphasize that they are the same size. Explain that when a whole is divided into two parts that are the same size, we call each part a half. Ask your child what the two parts together make. Help him or her to understand that together the two halves make a whole.

You may decide to introduce your child to thirds and quarters immediately or wait a while, depending on how readily your child understood the concept of dividing a whole into two halves. Whenever you decide to proceed, introduce the concept of dividing wholes into thirds and quarters the same way as you did for halves. Ask your child to count how many parts there are. Note that each part is the same size. Discuss how putting the parts together makes a whole. Place an emphasis on the name of the fraction that each part represents. You will likely have to revisit fractions many times before your child will be able to identify each type of fraction and tell you how many of them make up a whole.

Play dough also makes a good teaching aid for helping your child learn about fractions. Have your child roll out a play dough "snake" or other shape and then divide it into half's, thirds, or quarters.

As well, you may reinforce the concept of dividing wholes into fractions back in the kitchen. Cut fruit into two halves and then four quarters. Divide a bunch of grapes equally into three bowls. Cut your child's sandwich into two halves and then four quarters. Gradually fill a glass with milk. Note when the glass is a quarter full, a third full, half full and a full cup. As your child drinks the milk, note when he or she has drank a quarter of the milk, a third of the milk, half the milk and the whole glass. If you are eating a pizza, cut the pizza into two halves and then four quarters. Ask your child whether it is possible to make smaller fractions. Ask him or her to guess how many pieces there would be if you cut all of the quarters in half again? What would be a good name for such fractions?

The concept of fractions and the concept of sharing go well together. Ask your child what fraction to divide an apple, a pizza, or desert into in order to share it between family members.

Money Matters - Coins

36

Children learn to appreciate early on that there is something special about money. They hear people talk about money and they see how people treat money differently from other objects around the house. It doesn't take them long to figure out that money is used to buy things at the store. So money acquires a special allure even to young children. Coins are particularly attractive to children because they are heavy and shinny. Take advantage of that allure to teach counting and number concepts to your child.

Very early on, pennies make good counting objects for children. Once your child is learning to add and subtract, coins can be used effectively to add interest and variety to math sessions. Keep a bowl or jar of pennies, nickels, dimes, and quarters handy. Begin by helping your child to learn the names and values of the various coins.

Coins are very good for teaching the idea of equivalence, for example that five pennies are equivalent to one nickel in value, and that two nickels are equivalent to one dime in value. Start off by teaching your child how many pennies are "in" nickel, "in" a dime, and "in" a quarter. Put a nickel, dime, or quarter in front of your child and ask him or her to count out the equivalent number of pennies. Alternatively, count out five, ten, or twenty-five pennies, and ask your child to tell you which other coin is equivalent in value.

Once your child has a good understanding of the value of each type of coin, use coins to practice adding and subtracting. Coins are good to work with because you can quickly and easily make up addition and subtraction problems for your child to do. As well, it is simple to vary the difficulty of the exercise to match your child's ability simply by changing the number and type of coins that you set out.

Start by giving your child a small number of coins to count, such as one nickel and two pennies. Although your child may be able to tell you how many pennies are "in" a nickel, adding a nickel and two pennies together may very well cause confusion. For your child, a logical answer to the question, "how much does one nickel plus two pennies make" is "three".

When counting out addition questions, young children typically want to start counting at number one. It is difficult for children to start counting at a larger number such as five. You can help your child when he or she is adding together coins by suggesting that they tap out the value of the coin as they count out the answer. For example, tap a nickel five times and tap a dime ten times. Eventually, your child will be able to add coins without counting out the value. As your child's adding ability progresses, gradually give your child more and larger coins to count.

Coins can also be used effectively to provide practice subtracting. Give your child a number of coins and have him or her determine their total value. Then move one or more coins aside. First, ask your child tell you the value of the coins that were moved aside. Then ask him or her to calculate the value of the remaining coins. Younger children will have to count out the remaining coins as they are still learning the concept of subtraction. Older children should be able to determine the value of the remaining coins without counting all of them out.

The following "banking game" will help your child learn the relative value of different types of coins. You are the "bank" and as such will need a supply of pennies, nickels, dimes, and quarters. Have your child roll a single die and then give him or her as many pennies as the number he or she rolled. For example, if your child rolls a "3" give him or her three pennies. Then your child gets to roll again and collect more pennies. Whenever your child can exchange a number of coins for one larger coin he or she must do so. For example, if your child has eight pennies he or she must exchange five of them for a nickel from the bank. If your child has two nickels and four pennies, he or she must exchange the two nickels for a dime from the bank. Choose a target number, such as twenty-five or fifty, for your child to aim for in order to "win" the game.

CHAPTER THIRTY-SEVEN

Money Matters - Playing Store

A pleasant way for you and your child to practice adding and subtracting numbers is to play store. Begin by having your child select a number of toys or other items to "sell" in his or her store. He or she may also clip pictures out of catalogues and magazines to "sell" in the store. Selecting the pictures to clip can be a fun activity in itself. Help your child to decide on the price of each article being offered for sale. If you want to, make up little price tags to stick onto the articles with tape.

Choose something that you wish to purchase and ask your child what the price is. Then hand your child an amount in coins that is equal to or greater than the price of the article that you are purchasing. The first thing for your child to do is to add up how much money he or she received in payment. Next, he or she will have to determine how much change they owe to you, if any.

MARSHMALLOW MATH

Younger children can count out the difference between the sale price and the amount you paid. Older children can practice subtracting the sale price from the amount of money you paid. In either case, you control the difficulty level when setting the price of the articles put up for sale and when deciding on the amount of money that you will tender in payment. You may also increase the difficulty level by purchasing several items and asking your child determine the total sale price.

CHAPTER THIRTY-EIGHT

Estimating

38

While math may seem to be all about being precise and accurate, there isn't always the need or the time to be exact. When a journalist reports that 10,000 protesters marched on city hall, you can be sure that he or she did not count all of the protestors one by one. When conservation officers determine fish stocks in a lake, they most certainly do not count each and every fish. Both the journalist and the conservation officer are at best estimating the numbers of protesters and fish.

Even when you do want to arrive at a precise and accurate number, estimating is a useful exercise. Estimating the correct answer to a math question before doing the precise calculations is a good way of ensuring that your calculations are logical. Many math teachers have been shocked and amused by the wildly

wrong answers students give to math questions. A little common sense and a quick estimate of the approximate answer would have made the students realize that their calculations could not possibly be correct. The over reliance on calculators today has probably made this problem more common as students implicitly trust the answers given by a calculator. Calculators never make a mistake! The problem, of course, is that to obtain the correct answer using a calculator the person using it must input the correct information. As the saying goes; garbage in – garbage out. If a person cannot estimate what the correct answer should be, he or she will not know if they have made a mistake.

Your child can learn what it means to estimate and gain some skill at it by estimating the number of things. Place a fairly large number of counting objects, such as macaroni noodles, in a pile. Ask your child to guess at how many noodles there are in the pile. Then actually count out the noodles together. How close was the guess? You can help your child gain expertise by having them consider how large a pile ten noodles is and then use that as a reference for estimating how many are in the larger pile. Like any other skill, a person gets better at estimating with practice.

Another estimating activity is to have your child take a guess at how many bites it will take him or her to eat a particular food item. Have him or her estimate, for example, how many bites it will take to eat an apple, a bowl of cereal, or a slice of pizza. Similarly, he or she could estimate how many sips will it take to drink a glass of milk or juice? Obviously, your child will have some control over this so don't place a large bet on the outcome. However, the mere fact that your child can control the size of each bite or sip will help him or her to recognize the importance of "units of measurement" in estimating and quantifying.

When your child tackles other math questions and problems, occasionally suggest that he or she try to estimate the answer before working out the exact answer. The way to estimate answers is to round off to numbers that can be more easily calculated mentally. For example, if your child wants to add 16 and 23 together, he or she may quickly estimate that the answer will be close to "15 + 25 = 40". Help your child learn to make reasonable estimates. Ask him or her how they arrived at their estimate. If you can think of a better way of quickly estimating the answer, explain how you would go about it.

CHAPTER THIRTY-NINE

Measuring Up

39

Measuring things is a good way to encourage your child to use numbers. It is also a good way to reinforce the important concept of dimension by measuring the length, height, and width of objects.

Children will enjoy measuring things. Measure things such as books, toys, furniture, pets, and family members. Accuracy is not a big issue at this stage. Your child is not building a bridge!

With very young children, it is a common practice to introduce measuring using what may be called "nonstandard" units of measurement. That is, you can measure things using whatever unit of measurement you want. For small objects you might use things such as macaroni noodles, toothpicks, or paperclip chains to measure with. For example, you might count out how many paperclips long a small toy is. For larger objects, you might use the width of your child's hand or the length of your child's feet as the unit of measurement. Another approach is to first measure the height or length of the object using a string and then count

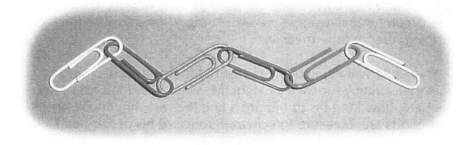

out how many hand widths or feet lengths the string is. The use of nonstandard units of measurement may be taken one step further by combining large and small units of measurement. A favourite toy might be, for example, six hand widths and four paperclips long.

If your child can read numbers provide him or her with a ruler or measuring tape to try measuring with. Begin by making your child familiar with the units of measurement. In Canada, schools teach measurement in meters and centimeters. In the United States, children are primarily taught to measure using feet and inches. Use whichever unit of measurement you prefer or both. Ultimately, you child should understand that there are different units of measurement and he or she should be comfortable using both of them. You may have to help your child read larger numbers.

When your child measures something, help him or her to determine whether he or she is measuring the length, height, or width of the object. Measure all three dimensions. Ask your child to compare the different dimensions of the object. Is it longer than it is tall? Is it taller than it is wide? Compare the measurements of one object to the measurements of another. Determine which objects are the longest, the tallest, and the widest.

For round and curved objects, such as balls or pots, measure the circumference using a piece of string. Then measure the string using a ruler or tape measure.

Your child will enjoy being measured themselves. Go beyond just measuring your child's height. Measure how long his or her feet are, how long his or her arms and legs are, how wide his or her hands are. Measure the circumference of his or her head, chest, and waist. Record such measurements and compare them to other family members. Allow your child to measure you.

After you have introduced the idea of estimating, as discussed in Chapter 38, ask your child to estimate an object's length, height, or width before measuring it. You may wish to record both the estimate and the actual measurement and then calculate the difference between them. To make this activity into more of a game, both you and your child can record your estimates and then determine who was closest. See who improves the most with practice.

CHAPTER FORTY

Graphic Fun

40

Hopefully, your child is enjoying learning to count, add, and subtract. However, much of what people do with numbers in the real world is to gather information or data, such as measurements or statistics, and present it to others in a user-friendly format. A common method of displaying such information is in graphs such as bar graphs and line graphs. Your child will get a kick out of conducting an inquiry, collecting data, and putting such data into a graph. Doing so will help your child appreciate how numbers can help us to better understand our world.

Approach this activity in a scientific manner. Start by formulating a question that you and your child want answered. An example might be, "Who in the family watches the most television?" Determine what data will be required to answer the question and decide on the best way to collect it. Then collect the required data and record it on a graph. The final step of your inquiry would be to analyze the graph to answer your original question.

What sort of inquiries might you and your child conduct? You might want to know whether your child has more winter clothing or summer clothing. You might want to take an inventory in the kitchen. How many pots and pans, cups, glasses, and plates do you have? You might want to determine how many fruits and vegetables each family member eats or how much milk or juice different family members consume. You might make inquiries about the weather. Is it usually warmer in the morning or in the afternoon? Are daily temperatures getting warmer or cooler as the season changes?

In Chapter 39, it was suggested that measuring was a good learning activity for your child. Recording such measurements in a graph is a good way of presenting the information obtained. One idea is to measure each family member's height or weight and record it in a graph. Another idea is to measure how far each family member can jump and record the distances on a graph.

You can purchase graph paper at any stationary store. As well, you may photocopy the graph paper found in the appendix of this book. Alternatively, you may use pictures or stickers to make pictographs. The following is a simple example of a graph that you might make with your child.

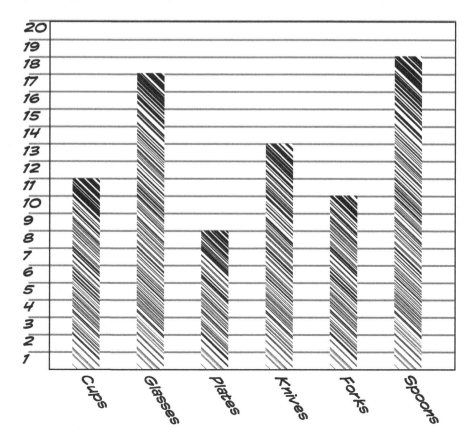

PART FOUR

Extreme Math

The following chapters go well beyond what is generally expected of grade one students. The reason for including the following chapters in this book is because the same approach to teaching basic concepts, using counting objects such as macaroni noodles, pennies, or jellybeans works so very well when teaching more advanced concepts. Once your child has mastered basic skills, he or she may not find the following concepts difficult to learn and will most likely enjoy doing so. Each skill builds incrementally upon earlier skills and so mastering the following skills will not be a big step for your child. While multiplying is a far more sophisticated skill than counting, learning to multiply may actually be easier for your child to learn. This is because he or she has already had considerable experience with numbers and has developed an ability to use them.

CHAPTER FORTY-ONE

Odd and Even

41

Knowing whether a number is odd or even tells us whether it can be divided by two evenly. It is a useful number concept and one that young children can learn.

Essentially, any number that can be reached by counting by two's from zero is an even number. Two, four, six, eight, ten, twelve, etc. are all even numbers. The numbers in between are all odd numbers. One, three, five, seven, nine, eleven, etc. are all odd numbers.

Introduce this concept to children using pennies or other counting objects. Count out an even number of pennies such as eight. Ask your child to count out the pennies in pairs of two in two neat rows. Ask him or her to look carefully at each row. Is there the same number of pennies in each row? Explain that there is the same number of pennies in each row because it is an even number. Then count out an odd number of pennies such as nine. Again, ask your child to count out the pennies in pairs of two in two neat rows. Is there one extra penny this time? Explain that there is an

MARSHMALLOW MATH

extra penny because nine is an odd number and cannot be divided evenly into pairs of two. With odd numbers there will always be one extra penny.

Do the above activity with all of the numbers from one to twenty. Have your child try guessing which numbers will turn out to be even and which will turn out to be odd. If your child is good at counting by two's, he or she will soon be able to identify even large numbers as being odd or even. The pattern is as follows. Numbers ending in a 0, 2, 4, 6, or 8 are all even. Numbers ending in a 1, 3, 5, 7 or 9 are all odd.

CHAPTER FORTY-TWO

More Fun with Fractions

42

Once your child is able to identify each type of fraction and tell you how many of them make up a whole, you can introduce the concept of adding and subtracting fractions. With the fraction circles available to look at and count, it is not as difficult for your child to do as it may sound.

Start by asking your child to identify a half circle. Then ask your child what he or she would get if they were to add another half circle to it. Put the two half circles together and you get a whole. Emphasis that one half plus one half makes a whole. Next, ask your child what you get if you subtract a half from a whole. Take away one half circle and the answer is obvious, you are left with one half.

Take the same approach with thirds. Ask your child to identify a third of a circle. Then put another third of a circle next to the first one. Ask your child how much of a whole circle you have. Your child will not know the answer, so help him or her to count out the answer. Very simply, you get "two-thirds". Add one more third of a circle and you get "three-thirds". Emphasize that three-thirds makes a whole. Continue by asking what would you get if you subtract one-third from a whole. You get two-thirds again. And if you subtract one-third from the two-thirds, you are left with one-third. Repeat the activity once or twice so that your child becomes familiar with the idea of counting the number of thirds.

Take the same approach to adding and subtracting quarters. Have your child count out the number of quarters so as to become familiar with the concepts of one-quarter, two-quarters, and three-quarters. Help your child to understand that two-quarters is equal to one half, just as four quarters is equal to a whole.

The next step, once your child is ready, is to add fractions to wholes. For example, take a whole circle and place a third of a circle next to it. What number do you get? You get one and one-third. Add another third and you get one and two-thirds. Explore all of the various combinations of wholes and fractions. Continue adding fractions as you build two wholes, three wholes, and even four wholes.

At this stage, it is too early to expect your child to add together fractions with different denominators. For example, you would not ask your child to add together "one-third plus one-quarter".

To take your child's understanding of fractions a little further, however, try a little subtraction. Start with two whole circles made up of, say, a whole circle and the four quarter circles. Take away one of the quarter circles and count out the answer. You are left with one and three-quarters. Continue to subtract quarters and count out the answer. Repeat the process with different fractions.

You will know that fractions have become a part of your child's vocabulary when they start asking for half an apple or tell you that they have eaten a third of their peas.

CHAPTER FORTY-THREE

Dozens and Dozens 43

With the world going metric, the concept of a "dozen" and counting by twelve's is probably declining in importance. However, it appears that eggs are still being sold by the dozen so it is probably still worthwhile knowing how much a dozen is.

The most useful teaching aid for helping children learn about dozens is, of course, the egg carton. If you trust your child, allow him or her to count how many eggs there are in a full egg carton. If the risk of broken eggs is too great, save an empty egg carton and have your child count while placing other less delicate objects, such as jellybeans, into the cups.

When your child is able to add together larger numbers, use the egg carton again to help him or her learn to count by twelve's. Have your child place two jellybeans in each cup and then count how many jellybeans there are all together. Next, place three jellybeans in each cup and count them again. Continue this activity as far as you and your child want to go.

CHAPTER FORTY-FOUR

The Secret Formula 44

This book places much emphasis on using counting objects and mental math as opposed to doing written work. This reflects two basic ideas. The first is that using counting objects will help to make abstract ideas more concrete for your child. The second is that counting and otherwise manipulating numbers mentally will stimulate your child's brain and enhance his or her capacity to learn and understand math. However, that is not to say that young children should never use paper and pencil to learn and do math.

While math mostly involves the manipulation of numbers, there is also a written language to learn. Your child will need to be able to recognize common math symbols such as the symbols for adding, subtracting, multiplying, and dividing. As well, your child will need to be able to solve simple equations.

Most of the time, when posing questions for our children, the equations given take the following common form: "given number plus given number equals unknown number". For example, "$3 + 4 =$ __". However, it is good for children to be able to read equations and understand what is expected of them when the unknown number is placed elsewhere in the equation. Your child may have difficulty turning an equation such as "$3 + 4 = 7$" around and looking at it from a different perspectives. For example, your child may not readily perceive that the question "$3 +$ __ $= 7$" is a variation of the equation "$3 + 4 = 7$". There is a tendency to assume that the unknown number is 10.

The following addition questions are all based on variations of the equation "3 + 4 = 7".

$$3 + 4 = \underline{}$$
$$4 + 3 = \underline{}$$
$$\underline{} + 4 = 7$$
$$\underline{} + 3 = 7$$
$$4 + \underline{} = 7$$
$$3 + \underline{} = 7$$
$$\underline{} = 3 + 4$$
$$\underline{} = 4 + 3$$
$$7 = 3 + \underline{}$$
$$7 = 4 + \underline{}$$
$$7 = \underline{} + 3$$
$$7 = \underline{} + 4$$

The following subtraction questions are all based on variations of the equation "7 - 4 = 3".

$$7 - 4 = \underline{}$$
$$7 - 3 = \underline{}$$
$$7 - \underline{} = 4$$
$$7 - \underline{} = 3$$
$$\underline{} - 3 = 4$$
$$\underline{} - 4 = 3$$
$$4 = 7 - \underline{}$$
$$3 = 7 - \underline{}$$
$$4 = \underline{} - 3$$
$$3 = \underline{} - 4$$
$$\underline{} = 7 - 4$$
$$\underline{} = 7 - 3$$

Children who can answer a question stated in the usual order can easily become confused by putting essentially the same question in a different order. To help your child learn to read and interpret equations, take a simple equation that your child comprehends well and rewrite all of the variations following the above patterns. Ask your child determine the unknown in each question. Repeat the activity with different equations until your child finds it easy to do. At first, allow your child to look at the completed equation, for example "3 + 4 = 7", to figure out what the unknown is in each variation. Once your child becomes confident at determining the unknown with a completed equation to refer to, ask him or her work on similar questions without having a completed equation to look at. Eventually, your child will be able to work out the unknown for any variation of a simple equation. At this point in time, it is probably too early to teach your child how to formally "solve" an equation to determine the unknown. He or she will learn to do so intuitively.

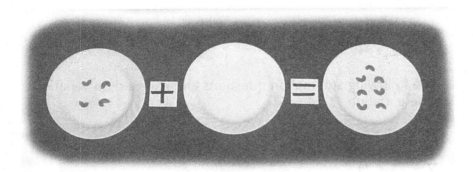

You may find it helpful to use counting objects, such as macaroni noodles, to help make the above variations of an equation more concrete for your child. Use three small saucers to put the noodles in. On three separate pieces of paper, print a large " + " symbol, a large " – " symbol, and a large " = " symbol. Recreate the various equations using the saucers, symbols, and noodles by leaving one saucer empty as the "unknown".

CHAPTER FORTY-FIVE

Multiplication for Beginners

It is good to keep in mind that a distinction can be made between understanding a concept and being skilful at using the concept to answer questions or solve problems. Most often the two abilities go together so that someone who understands a concept can also use it skillfully to solve problems. However, it is possible to gain a good understanding about a mathematical concept, but not yet be skillful at using the concept to solve problems or answer questions. The skill comes from practice and from memorizing number facts. The reverse is also possible. A child may be trained to do a certain type of calculation well without understanding the underlying mathematical principles.

The concept of multiplication provides a good example of both possibilities. A child may be taught to memorize the multiplication table without gaining a sound understanding of what it really means to multiple two numbers together. Such a child may be able to quickly recite the answer to the question, "how much is four times eight", but not understand why the answer is "32". Such rote memorization is of little value in becoming proficient in mathematics. On the other hand, a child may have a very good understanding of what it means to multiple two numbers together, but not have yet memorized all of the number facts found in the multiplication table. While that child is not yet skillful, he or she is well on their way to mastering multiplication. Proficiency will come, in time, with the memorization of multiplication number facts.

Your first goal with your child, therefore, is to help him or her gain a firm idea of what it means to multiple two numbers together. Explain that multiplication is simply a form of addition. It is adding the same number over and over again a certain number of times. For example, "three times two" is the same as "two plus two plus two". The answer in both cases is "six". Fortunately, the concept is very simple to illustrate using counting objects.

Your child will acquire a good idea of what multiplying means by doing the following activity. Using counting objects such as pennies, ask your child to show you what "three times two" looks like. Help your child to form three groups of pennies with two pennies in each group. Then ask your child to tell you how much "two plus two plus two" equals. Encourage your child to count out the answer. Next, ask your child how much "three times two" equals. The answer, of course, is the same. You have merely changed the form of the question. Even if your child wants to recite the answer immediately, encourage him or her to count out the answer again.

Taking a closer look at the above example, we can identify three basic steps. The first step is to have your child show you what multiplying  looks like by making groups of pennies. He or she has to determine many groups of pennies are required and how many pennies must be in each group? The second step is to add up the groups of pennies. The third step is to repeat the addition question as a multiplication question. This approach emphasizes that multiplying is a form of addition. Reinforce this by stating that "three times two" is the same as "two plus two plus two"

Repeat the above activity multiplying different numbers

together. Please remember, however, that at this point in time your child has not yet memorized multiplication number facts. Each time that your child is asked to multiply "four times three", he or she will have to add up the answer until it is memorized.

You will help your child by being consistent as to which number refers to the number of groups and which number refers to how many pennies there are in each group. For example, "four times three" should always mean "four groups of three pennies" and not "four pennies in each of three groups". Eventually, your child will know that the answer is the same. However, when your child is trying to determine how many piles of pennies to set out it is important to be consistent.

Don't forget the number zero. Explain to your child that zero times any number equals zero, and that any number times zero also equals zero. For example, "zero times four equals zero" and "four times zero also equals zero". For a little fun, ask your child to set out "zero groups of four pennies" or "four groups of zero pennies".

Blank number tiles are also useful to illustrate what multiplication looks like. Have your child arrange blank tiles into squares and rectangles to demonstrate what, for example, "four times three" or "four times four" looks like. Have your child count how many tiles there are along each side and how many tiles there are altogether.

Multiplication was invented because adding many numbers together becomes cumbersome. Likewise, working with many groups of pennies with many pennies in each group can quickly become unwieldy. Doing the following multiplication questions should be sufficient to provide your child with a very good understanding of what multiplication means. Larger numbers can wait until your child is ready to begin memorizing the multiplication table.

MARSHMALLOW MATH

One times one equals one
One times two equals two
One times three equals three
One times four equals four
One times five equals five
Two times one equals two
Two times two equals four
Two times three equals six
Two times four equals eight
Two times five equals ten
Three times one equals three
Three times two equals six
Three times three equals nine
Three times four equals twelve
Three times five equals fifteen
Four times one equals four
Four times two equals eight
Four times three equals twelve
Four times four equals sixteen
Four times five equals twenty
Five times one equals five
Five times two equals ten
Five times three equals fifteen
Five times four equals twenty
Five times five equals twenty-five

Division for Beginners

46

Once your child has a firm understanding about what it means to multiply two numbers together, it is appropriate to introduce the concept of division. There is no need to wait until your child has memorized the entire multiplication table. Division, of course, is simply the reverse of multiplying. We start with a larger number such as twenty and then try to determine what two numbers, (one given and one unknown), multiply together to produce it. If we divide twenty by the given number five, we can determine that the unknown number is four. (At this stage, only give your child division problems that you know will work out evenly.)

Fortunately, demonstrating the concept of division is simple using counting objects such as jellybeans. Start with a number of jellybeans, (which you know to be the product of two smaller numbers multiplied together), and ask your child to divide the jellybeans into a certain number of equal groups.

MARSHMALLOW MATH

For example, let's divide the number twelve by the given number three. Start by asking your child to count out twelve jellybeans. Then ask your child to divide the twelve jellybeans into three equal groups. Have your child begin by creating three groups by setting out three jellybeans. Next, have your child continue adding jellybeans one at a time to each group until all of the jellybeans are gone. There should be four jellybeans in each group. Explain to your child that "twelve divided by three equals four". For good measure, also point out to your child that "three times four equals twelve".

Take the same twelve jellybeans and ask your child to divide them into four equal groups. Again, have your child begin by creating four groups by setting out four jellybeans. Before he or she goes any further, you might ask your child to guess how many jellybeans will be in each group. Then have your child continue adding jellybeans one at a time to each group until all of the jellybeans are gone. There should, of course, be three jellybeans in each group. Reinforce the learning by stating that "twelve divided by four equals three" and that "four times three equals twelve".

Give your child practice dividing by repeating the above activity using other numbers. All of the multiplication questions listed in Chapter 45 also make suitable division problems at this stage.

Sharing candies or cookies among friends and family members provides a good opportunity to practice division. Count out how many treats there are and then ask your child to figure out how to divide them equally between everyone. If the number does not work out evenly it may also provide an opportunity to revisit fractions.

As your child begins to memorize multiplication facts, for example that "three times five equals fifteen", encourage him or her to also memorize the equivalent division facts. For example, that "fifteen divided by three equals five".

Memorizing Multiplication and Division Number Facts

47

If your child has demonstrated a good understanding of what it means to multiply two numbers together and has started memorizing some of the answers to multiplication questions, you may want to encourage him or her to put more effort into memorizing multiplication facts. Understanding what multiplication means is one thing, becoming skillful by memorizing the multiplication table is another thing.

There are two tried and true ways to help children memorize multiplication facts. One is the multiplication table and the other is a set of multiplication flash cards. Using both tools will assist your child in memorizing common multiplication facts, which will be of enormous benefit once he or she begins to learn more advanced mathematics.

One word of caution, avoid having your child memorize answers to multiplication questions that he or she cannot figure out independently. That is, do not ask your child to memorize that "seven times eight equals fifty-six", unless and until he or she can independently work out that "seven time eight equals fifty-six." The first step for your child should be to figure out what seven time eight equals using addition. That is, he or she should be able to add "8 + 8 + 8 + 8 + 8 + 8 + 8" together and arrive at the correct answer. The second step is to then memorize the answer so that he or she no longer has to do all of that addition each time.

This approach will help to ensure that there is real understanding backing up the memorization. As well, if your child were to forget the answer he or she will be able to work it out with confidence. Further, this approach provides a natural pace for memorizing multiplication facts and helps to avoid the pitfall of overwhelming your child.

If used wisely, multiplication flash cards are an excellent teaching aid for helping your child to memorize multiplication facts. You can either purchase a commercial set of cards or make up your own using 3-inch by 5-inch index cards. A multiplication question is printed on one side of the card and the answer is provided on the reverse side of the card, (usually in smaller print).

When using flash cards, only give your child cards for the multiplication questions that he or she has already worked out independently. It will only be a small pile of cards to begin with, but will gradually grow in size as he or she learns to multiply larger numbers. This way, your child will only use the cards to review multiplication questions that he or she has been able to calculate. If, while going through the cards, your child forgets an answer try to resist the temptation of allowing him or her to simply look on the reverse side for the answer. Instead, ask your child to calculate the answer again. It is best to only look on the reverse side to confirm and reinforce a correct answer. Yes, it is more work, but the mental effort will stimulate your child's brain and develop his or her math skills more than if he or she simply reads the answer. As well, having to work out the answer rather than simply reading it may help motivate your child to memorize the multiplication table more quickly.

MEMORIZING MULTIPLICATION AND DIVISION

In keeping with the above approach, do not just give your child a multiplication table to memorize. Instead, use a blank multiplication table on which to record the multiplication questions that your child has been able to memorize. He or she can then use the multiplication table for a quick review. Fill in the blank spaces slowly as your child gradually memorizes more and more of the multiplication table. The blank spaces in the table clearly indicate the learning that still has to take place and provides a great incentive for moving ahead. An example showing how to use a blank multiplication table is shown below. You may photocopy the blank multiplication table found in the appendix of this book for your child to use.

Multiplication Table

	1	2	3	4	5	6	7	8	9	10
1	1	2	3	4	5	6	7	8	9	10
2	2	4	6	8	10	12	14			20
3	3	6	9	12	15	18				30
4	4	8	12	16	20					
5	5	10	15	20	25					
6	6	12	18							
7	7	14								
8	8									
9	9									
10	10	20	30							

It is one thing to remember that "seven times nine equals sixty-three". It is another thing to remember that "sixty-three divided by nine equals seven". To be skillful at dividing, your child should also be able to recall such common division facts.

To a large extent, of course, the memorization of multiplication facts is also the memorization of division facts. However, it is helpful to focus on division. One simple way is to use your child's multiplication table the other way round. Choose any number found in the body of the table and ask your child what two numbers may be multiplied together to produce it. Remember that some numbers are the product of more than one combination of numbers. For example, the number 24 is the product of "1 x 24", "2 x 12", "3 x 8", and "4 x 6".

You might also consider making up division flash cards. Write division questions such as "24 ÷ 3 =" on one side of the card and the answers, in this case the number "8", on the reverse side of the card. The same considerations about using multiplication flash cards also apply, of course, to division flash cards.

CHAPTER FORTY-EIGHT

Place Value

48

By now, your child already has some awareness about the concept of place value. However, it is best to ensure he or she has a solid grasp of the concept before attempting to do addition or subtraction questions with larger numbers.

As you know, we have a base ten number system. As you may want to explain it to you child, it is worth a quick review here. Adding ten 1's together gives us the number ten. Adding ten 10's together gives us the number one hundred. Adding ten 100's together gives us one thousand and so on. It also means that once we run out of single digit numbers, (after the number 9), we then have a two digit number which is the number 10. After we run out of two digit numbers, (after the number 99), we then have a three digit number which is the number 100. And so on.

The place value of digits is important because each digit in a number represents a different order of magnitude. In the number 427, the digit 7 represents the number of 1's, the digit 2 represents the number of 10's, and the digit 4 represents the number of 100's. That is, the number 427 is made up of four 100's, two 10's, and seven 1's.

You can help your child understand this fundamental concept using macaroni noodles or other counting objects. As you might not want to illustrate numbers as large as 427 using noodles, try working with a more manageable number such as 36. Draw a line down the middle of a piece of paper. Tell your child that the right side of the line represents the number of 1's found in the

MARSHMALLOW MATH

number and that the left side represents the number of 10's found in the number. At the top of the page print the number 36, putting the digit 3 on the left side of the line and the digit 6 on the right side of the line. Ask your child to count out 36 macaroni noodles placing three groups of ten noodles on the left side of the line and six macaroni noodles on the right side of the line. Explain how the number 36 is made up of three 10's and six 1's. Pick several other double-digit numbers and have your child separate out the 10's and the 1's. Ask your child what he or she would have to do with a bigger number such as 427 or 3,765.

You may illustrate the same concept using dimes and pennies. Set out a certain number of dimes and pennies, for example six dimes and four pennies, and ask your child to tell you what number is represented. Alternatively, pick any number from 0 to 100 and ask you child to make up the number using dimes and pennies.

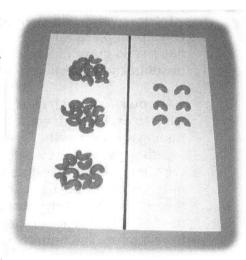

Reinforce the concept of place value using number tiles. Use number tiles 0 through 9. Give your child any two numbers and ask him or her to make up the largest two-digit number possible using those two numbers. Next, ask him or her to make up the smallest two-digit number possible using those same two numbers. Repeat this activity with various numbers. Have your child say out loud each number that they make up.

Once your child is competent using two number tiles, give him or her three number tiles to work with. What are the largest and the smallest three-digit numbers that he or she can make with the three number tiles? Ask him or her which digit represents the number of 100's, which digit represents the number of 10's, and which digit represents the number of 1's. Try the same activity with four or five digit numbers.

CHAPTER FORTY-NINE

Adding Huge Numbers

Once your child can add together any two numbers from zero to ten with confidence and understands the concept of place value, he or she is very close to mastering the addition of whole numbers altogether. There are really only two ways to make the addition of whole numbers more difficult. One way is to add more numbers together. The other way is to add bigger numbers together. However, no matter how many numbers are added together or how big the numbers are, it always comes down to adding together the numbers from zero to ten.

Adding more numbers together or adding larger numbers together is not really more difficult than adding together two single digit numbers; for the most part it is simply more time consuming. As well, the opportunity for error increases because what you are doing in fact is adding many times to arrive at the answer to one question. Both factors, the time factor and the error factor, can take the joy out of doing such questions for many people. This is one reason why we have invented calculators. However, if you and your child have progressed this far and have kept math sessions short and fun, there is no reason for not taking the next step.

Remember that there is no need to do hundreds of problems. Doing several large addition problems once in while is an enjoyable challenge. Doing pages and pages of tedious problems is only work. Remember also that your child does not have to work out the questions entirely on his or her own. There is no rule that two people cannot work together to solve a problem. It is called cooperation. Over time, your child will require less and less assistance from you. You may be surprised at how much pride your child will feel after adding together huge numbers like "2,374,978 plus 4,284,385".

Introduce the addition of larger numbers using counting objects such as pennies. Use a piece of paper with a line drawn down the middle of it. Begin by adding together two double-digit numbers such as "28 + 14". Have your child count out the number of 10's and the number of 1's for each number in pennies on the lined paper leaving a space between the two numbers. Then ask your child to add together all of the pennies in the 1's column. As the total of twelve is larger

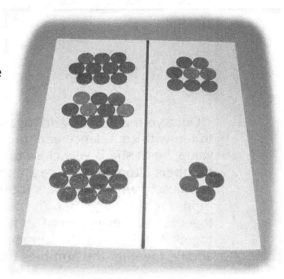

than ten, explain that he or she will have to "carry" ten of the pennies over into the 10's column leaving only two pennies in the 1's column. Next, ask your child to count how many piles of ten there are in the 10's column including the ten pennies that were carried over. The total is forty pennies. The answer to the question, of course, is forty-two.

The important thing is that your child truly understands the importance and implication of place value and does not just try to memorize "rules" about "carrying". Trying to do math by following "rules" can lead to many mistakes.

Once your child is able to add numbers together that require a "carry over" using counting objects, you may want to show your child how to do the same thing using a pencil and paper. For example:

$$
\begin{array}{r}
28 \\
+\ 14 \\
\end{array}
\qquad
\begin{array}{r}
{}^{1\ 0} \\
28 \\
+\ 14 \\
\hline
2 \\
\end{array}
\qquad
\begin{array}{r}
{}^{1\ 0} \\
28 \\
+\ 14 \\
\hline
42 \\
\end{array}
$$

Make up another similar questions and work them out using paper and pencil. Emphasize that we always start adding 1's together first, then the 10's, then the 100', etc. Once your child is competent at adding together two double-digit numbers, encourage him or her to try to adding together three or four double-digit numbers.

Adding together very large numbers such as "4,695 plus 3,568" is difficult to illustrate using macaroni noodles, pennies, or jellybeans. However, once your child understands that each digit represents a different order of magnitude, it is not difficult to work through even very large numbers together. Try working through the following question with your child.

$$
\begin{array}{r}
4,695 \\
+3,568 \\
\end{array}
$$

In the 1's column, "5 plus 8 equals 13". Carry the extra 10 over to the 10's column and print the digit 3 in the 1's column.

$$
\begin{array}{r}
{}^{1\ 0} \\
4,695 \\
+3,568 \\
\hline
3 \\
\end{array}
$$

In the 10's column, "90 plus 60 plus the carried 10 equals 160". Carry the extra 100 over to the 100's column and print the digit 6 in the 10's column.

$$
\begin{array}{r}
{}^{1\ 1\ 0} \\
4,695 \\
+3,568 \\
\hline
63 \\
\end{array}
$$

MARSHMALLOW MATH

In the 100's column, "600 plus 500 plus the carried 100 equals 1,200". Carry the extra 1,000 over to the 1,000's column and print the digit 2 in the 100's column.

$$\begin{array}{r} {\scriptstyle 1\,1\,1\,0} \\ 4{,}695 \\ +3{,}568 \\ \hline 263 \end{array}$$

In the 1,000's column, "4,000 plus 3,000 plus the carried 1,000 equals 8,000". Print the digit 8 in the 1,000's column.

$$\begin{array}{r} {\scriptstyle 1\,1\,1\,0} \\ 4{,}695 \\ +3{,}568 \\ \hline 8{,}263 \end{array}$$

There is no doubt that adding together lots of numbers and adding together big numbers takes time and effort. The effort will be rewarded with increased skill and confidence in math.

CHAPTER FIFTY

Subtracting Huge Numbers

50

Once your child understands the place value of digits in numbers and is able to add together larger numbers that involve a "carry-over", you may introduce similar subtraction problems. Subtraction is simply the reverse of addition. The equivalent to the "carry-over" in subtraction may be called "borrowing". If the digit being subtracted is larger than the digit it is being subtracted from, it is necessary to "borrow" from the next higher place value digit.

This concept of borrowing may also be demonstrated using counting objects such as jellybeans. Again, use a piece of paper with a line drawn down the middle. Let's look at the subtraction problem "24 – 6". The problem for your child is that 6 is larger than 4, so how can 6 be subtracted from 4? It would be possible, of course, to simply count backwards from 24 to determine the answer, but this would become rather cumbersome with larger numbers. Instead, explore the concept of place value to solve the question.

Count out 24 jellybeans on the lined paper. Place two piles of ten jellybeans in the 10's column and place four jellybeans in the 1's column. Ask your child whether the number 6 is bigger or smaller than the number 24. Point out that because the number 6 is smaller than the number 24, it should be possible to subtract 6 from 24. Then ask your child if it is possible to subtract 6 from 4. He or she will likely say that it is not possible. Suggest that it would be possible to "borrow" 10 jellybeans from the 10's column and add them to the 4 jellybeans in the 1's column.

MARSHMALLOW MATH

Move one group of ten jellybeans over to the group of 4
jellybeans making one large group of 14 jellybeans. Now ask your
child to take away 6 jellybeans from the group of 14 jellybeans.
The result is that there is now only one group of 10 jellybeans in
the 10's column and 8 jellybeans in the 1's column, for a balance
of 18 jellybeans.

Repeat the above activity with different numbers until your child
has a firm grasp of the concept of borrowing. Vary the size of the
numbers so that borrowing is sometimes required and
sometimes not required.

Once your child can work through such problems with
confidence using counting objects, you may want to show him or
her how to do the same thing using a pencil and paper. For
example:

$$
\begin{array}{r} 24 \\ -\,6 \\ \hline \end{array}
\qquad
\begin{array}{r} {}^{1}2^{1}4 \\ -\,6 \\ \hline 8 \end{array}
\qquad
\begin{array}{r} {}^{1}2^{1}4 \\ -\,6 \\ \hline 18 \end{array}
$$

The concept of borrowing from the higher place value digit
works with numbers of any size. Let's work though a larger
example, subtracting 268 from 534.

$$
\begin{array}{r} 5\ 3\ 6 \\ -2\ 6\ 8 \\ \hline \end{array}
$$

Begin by asking your child which of the two numbers is larger
and which is smaller. Point out that because 268 is a smaller
number than 536, it should be possible to subtract 268 from 536.
Ask your child to suggest how he or she could subtract the
number 8 from the number 6. The solution is to "borrow" 10
from the 30 found in the 10's column for a total of 16 in the 1's
column. Subtracting 8 from 16 leaves 8. Print the digit 8 in the
1's column. Because we borrowed 10 from the 30 in the 10's
column, there is now only 20 in the 10's column. Scratch out the
digit 3 and write in the digit 2 above it as a reminder.

$$\begin{array}{r} 2 \\ 5\;\cancel{3}\,{}^{1}6 \\ -\;2\;6\;8 \\ \hline 8 \end{array}$$

Next, ask your child how he or she can subtract 60 from the 20 in the 10's column. The solution is to borrow 100 from the 500 found in the 100's column for a total of 120 in the 10's column. Subtracting 60 from 120 leaves 60. Print the digit 6 in the 10's column. Because we borrowed 100 from the 500 in the 100's column, there is now only 400 in the 100's column. Scratch out the digit 5 and write in the digit 4 above it as a reminder.

$$\begin{array}{r} 4\;\;12 \\ \cancel{5}\;\cancel{3}\,{}^{1}6 \\ -\;2\;6\;8 \\ \hline 6\;8 \end{array}$$

The final step, of course, is to subtract the 200 from the 400 in the 100's column. Subtracting 200 from 400 leaves 200. Print the digit 2 in the 100's column.

$$\begin{array}{r} 4\;\;12 \\ \cancel{5}\;\cancel{3}\,{}^{1}6 \\ -\;2\;6\;8 \\ \hline 2\;6\;8 \end{array}$$

Make up similar two-digit and three-digit subtraction questions for your child to solve. A couple of challenging subtraction questions during each math session is sufficient to develop and reinforce subtraction skills. If you feel ambitious, try working through the subtraction of very large numbers, such as "5,738,291 – 3,465,325" with your child to demonstrate how the same principles apply to numbers of any size. He or she will likely be thrilled to be working with numbers in the millions.

Subtraction problems like the above are by no means easy for children to master. As calculating the answer involves numerous steps, there are many opportunities for your child to make an error. What is important is that your child develops a good comprehension of the place value of numbers and understands how to use that knowledge to solve questions. Remember that your primary goal is to help your child acquire the capacity to understand and use numbers. The mental effort exerted by your child will enhance his or her ability to learn and do more advanced math in the future.

MULTIPLICATION TABLE

	1	2	3	4	5	6	7	8	9	10
1										
2										
3										
4										
5										
6										
7										
8										
9										
10										

MARSHMALLOW MATH

GRAPH PAPER

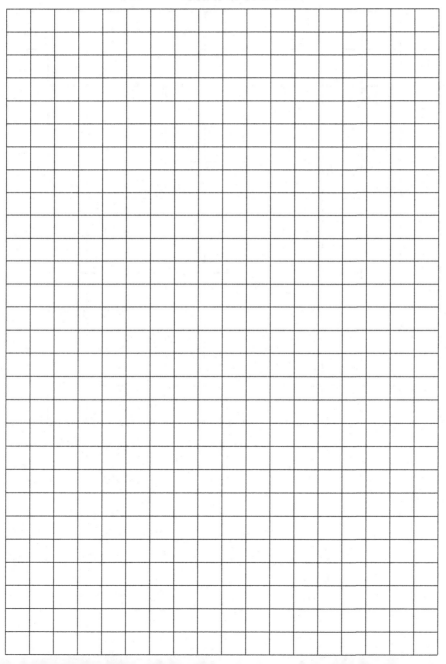

TANAGRAM

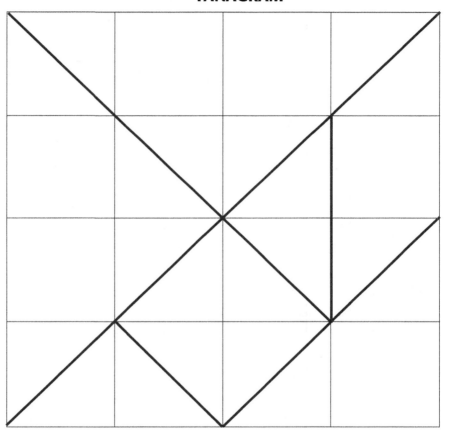

MARSHMALLOW MATH

HUNDRED NUMBER CHART

1	2	3	4	5	6	7	8	9	10
11	12	13	14	15	16	17	18	19	20
21	22	23	24	25	26	27	28	29	30
31	32	33	34	35	36	37	38	39	40
41	42	43	44	45	46	47	48	49	50
51	52	53	54	55	56	57	58	59	60
61	62	63	64	65	66	67	68	69	70
71	72	73	74	75	76	77	78	79	80
81	82	83	84	85	86	87	88	89	90
91	92	93	94	95	96	97	98	99	100

BLANK HUNDRED NUMBER CHART

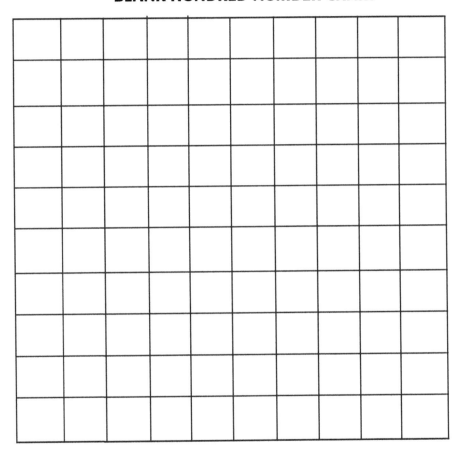